OUTDOOR OPTICS

OUTDOOR **Optics**

LEIF J. ROBINSON

LYONS & BURFORD, *PUBLISHERS*

Printed in the United States of America

10   9   8   7   6   5   4   3   2

Library of Congress Cataloging-in-Publication Data

Robinson, Leif J.
    Outdoor optics / Leif J. Robinson.
        p.   cm.
    ISBN 1-55821-065-2 : $13.95
    1. Binoculars.   2. Binoculars—Recreational use.
3. Bird watching—Equipment and supplies.   I. Title.
QC373.B55R63   1990
681'.412—dc20   *89-13770*
                                                           CIP

Contents

To Caroline,

who keeps the hearth warm while I seek the wonderment and diversity of Nature.

OUTDOOR OPTICS

Preface:
How to Go Wrong

For over twenty years I've spied upon bird-watchers almost as often as the birds themselves. At first my motive was to check out the various types of binoculars bird-watchers carried afield so that I'd know what to buy when I had to replace my beat-up hand-me-down pair. But the more I observed and the more I learned about how binoculars are made and how they form images, the more it became evident that most people carried models that were not the best for bird-watching.

In the 1960s many bird-watchers toted 7 × 50s, pronounced "seven by fifties." These binoculars have front lenses, called *objectives*, 50 mm (2 in.) in diameter, and they produce an image seven times larger than what one sees with the unaided eye. But none of my friends could give convincing reasons why they selected that particular type of binocular. "Outdoor Joe likes 'em!" was the usual response. Later I learned that 7 × 50s had been especially designed for use during World War II because they give the brightest possible image for sailors standing a night watch. That asset, not surprisingly, is of little importance to bird-watchers!

As I became increasingly aware of binocular design, function, and use, I began to have less and less confidence in the opinions of my friends. It became evident that a "knowledgeable" person was merely passing along his or her arbitrary choice to someone else. Bird-watchers, I uncharitably concluded, just didn't know much about their most important piece of field equipment.

If bird-watchers were having problems selecting an appropriate binocular, it occurred to me that other outdoor people might be having an equally difficult time making informed choices. That's how *Outdoor Optics* came to be. The situation today is especially complicated, for there exists a plethora of binocular and spotting-scope manufacturers, designs, and special features, all accompanied by prices that range from a few tens of dollars up to thousands. It is very easy to waste money on a piece of equipment that isn't of top quality or isn't matched to your needs.

As will become evident, no single model of binocular or spotting scope—no matter how good or popular—will be the best choice for everyone. Which one is best for looking at geological outcroppings during a backpacking trip? Which one brings butterflies really close up? Which one will be most convenient for watching shorebirds? Which one will give truly good views of features on the moon? In this book I'll describe the functions—mechanical, optical, and human—that should be considered when selecting an instrument that is just right for *you*.

What you won't find covered are underwater optics (I hate to get wet!) and scopes for rifles and handguns (I don't object to shooting; I just don't know anything about the sport). Yet, even though these applications aren't described explicitly, there is a lot of general information in the text that should be of value to prospective purchasers of such equipment.

If, after reading this book, you would like to explore further the subject of binoculars, telescopes, and their uses, I recommend Henry E. Paul's *Binoculars and All Purpose Telescopes* (New York, NY Amphoto, 1980), and Robert J. and Elsa Reichert's *Binoculars and Scopes*, (Philadelphia, PA Chilton, 1961). Both are rather dated, but they touch on some topics not covered in this work.

1

Selecting a Good Binocular

Fine optical instruments are quite expensive; they have to be, for they are precision products. When the top line of some binocular ad screams "Magnifies twenty times! See up to fifty miles!" and the bottom line directs you to "Send $29.95 to Box PU, Forgetit, Idaho," you should immediately be on guard. Don't be misled by advertising claims extolling magnification. Many people seem to be obsessed with the concept of magnification; they believe high magnification (often called "power") means better quality. This is simply not true. High-power binoculars (10 power is very popular today among the bird-watching fraternity) often have superb optics, but the power itself does not guarantee first-rate performance. High

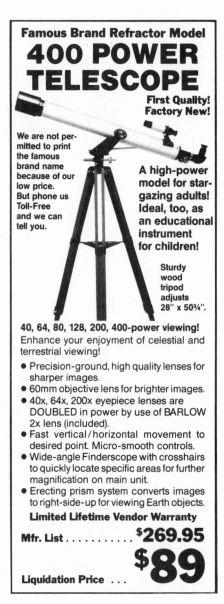
Ads need to be read with caution. The one containing this excerpt appeared in a national news magazine in 1986, shortly after the hoopla surrounding the appearance of Halley's Comet faded. The intent clearly seems to be to unload a manufacturer's unsold stock. Let's look at some of the ad's statements and implications. The brand name of the telescope is not stated, and high power is stressed. Also note the emphasis given to features that should be expected from *any* quality instrument: "Sturdy wood tripod," "precision-ground . . . lenses," "Wide-angle Finderscope with crosshairs," and so forth. And what, exactly, is a "Limited Lifetime Vendor Warranty"?

power will actually enhance the conspicuousness of any defects a binocular might have.

A good binocular will combine an appropriate amount of magnification with image clarity and brightness, and it will package those features in a framework that can withstand a good deal of knocking about. My message is simple: Never

By going to places where bird-watchers or other naturalists congregate, you can readily examine a wide variety of binoculars and spotting scopes. Here is a group of September hawk-watchers atop Wachusett Mountain in Princeton, Massachusetts. (Photo by author)

buy "bargain" binoculars from a manufacturer whose name you do not recognize as indicative of quality. Here are a few brands to remember: Bausch & Lomb, Fujinon, Leitz, Mirador, Nikon, Swift, and Zeiss. There are also many excellent companies that are known mainly in Britain and on the Continent, such as aus Jena, Steiner, and Swarovski. (Addresses for many firms are given in Appendix A.)

If you are just getting started, try any binocular that happens to be lying around the house — it's better to go afield with inferior equipment, whose faults should soon become apparent, than to go with nothing at all. If you can, compare carefully what you see in your binoculars and in those of your friends. Ask what features of their binoculars they especially like: magnification, image sharpness, brightness, field of view (that is, how wide a scene is visible), lightness, durability, "feel," price, manufacturer's service policy, and so forth. Compare as many models and brands as you can before buying your first binocular to be dedicated to any facet of nature study or other outdoor activity. Making that selection involves a tangle of factors, as you will see, though the basic qualities you should check out are easy to understand.

Humans differ. That statement is so obvious that its implications are often overlooked when selecting binoculars. Your physical size could be a prime consideration—an 85-pound person will not be able to comfortably support and carry a relatively heavy binocular (say, 30 oz or more). It is clearly in this person's best interests to select a lightweight model. Elderly persons are less likely to hold a binocular as steady as someone in his twenties; their best interests might be served by choosing a relatively low-power instrument. It is important to remember that any increase in magnification makes "jittering" due to hand tremor more apparent.

Compare the difference in bulk of these two excellent binoculars. On the left is the original 38-oz version of Swift's 8.5 × 44 Audubon (the latest version of this binocular weighs only 29 oz). On the right is the 16-oz Bausch & Lomb 7 × 24 Discoverer Compact, which, unlike its larger cousin, is waterproof. (Photo by author)

In addition to physical considerations, the manner in which the binocular will be used is also important. Is it to be taken out of its case only occasionally, say on sunny spring days or for viewing from the shelter of a living room, or is it going to be used (and abused) daily in all kinds of weather? Do you usually travel by car or do you climb mountains? If a binocular

Selecting a Good Binocular **9**

The Zeiss 10 × 40 binocular is depicted in *top*, unarmored and *bottom*, armored versions. The latter weighs nearly 9 percent more. (Photo courtesy Zeiss Optical Inc.)

At top is a pair of Nikon 7 × 35 binoculars and below it a pair of
Steiner 6 × 30. The Nikon instrument has all the obvious features of a
general-purpose binocular, such as center focusing and a conventional
body. The specialized Steiner, on the other hand, includes rubber
armoring, attached lens caps, and rain guard. (Photos courtesy Nikon
Inc., and Pioneer Marketing & Research, Inc.)

is to be used infrequently and handled with loving care, a rather flimsy body incorporating good optics would suffice. But if the binocular is going to get rough treatment or be exposed to rain or high humidity, a rugged water-resistant body is essential. Naturally, such durable construction is usually more expensive.

Yet ruggedness can be taken to an extreme. It is not uncommon to see weekend naturalists carrying "armored" binoculars, ones encased in a heavy rubber sheath. I believe many persons buy such equipment under the impression that the rubber waterproofs the instrument. It doesn't. Others may be merely satisfying their own macho images—after all, armored binoculars *were* developed for wartime use. Although outdoor activities sometimes parody a battlefield, they don't often make the same demands on your equipment! And, surely, there are people who buy a particular model simply because it's hawked by some well-known individual in an advertisement.

Before we get into details, I want to stress one easily overlooked fact. Unless you are really sure about what you intend to do with your equipment, and unless you know exactly what performance you want, buy something that isn't too specialized or isn't extremely expensive; in other words, get a Ford rather than a Maserati. By doing so, if your interest changes you'll still have a piece of gear that will serve you well, be it at the beach or the opera.

What Optics Do

Any binocular or spotting scope has two functions: to gather more light than the human eye does, and to enlarge the image of the subject. Before we explore how these instruments work, it is important to realize that the eye itself is an optical device. By using muscles and liquids, the eye works very much like the metal and glass that make up all optical instruments.

The pupil of your eye, and its response to various levels of illumination, largely determines why binoculars are made the way they are. When you look at a mirror, the pupil is the dark spot in the center of your eye. This little opening is the only place light can enter your eye, and it changes size according to the intensity of the illumination that strikes it. In fact, as

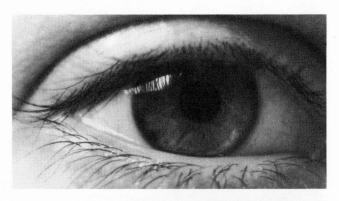

In twilight the pupil of a middle-aged person's eye will be about 4 mm in diameter, as in the top frame. (In bright sunlight the pupil's diameter can contract to as little as half this width.) The bottom frame shows the pupil's maximum dilation, about 8 mm. (Photos courtesy *Sky & Telescope* magazine)

described below, the maximum size your pupil can attain might determine why you choose one particular instrument over another.

In bright sunlight a typical pupil will be only 2 to 4 mm (¹⁄₁₂ to ¹⁄₆ in.) in diameter. In deep shade or at twilight the

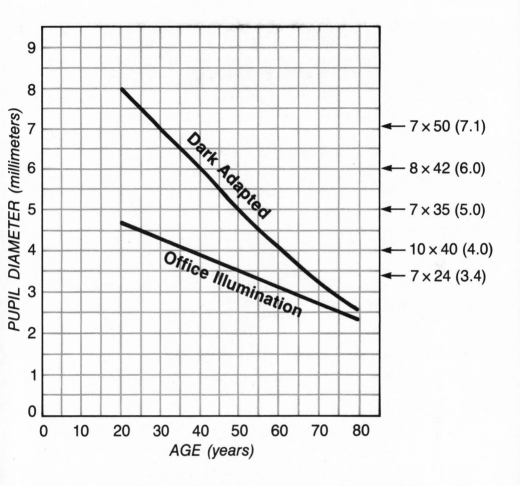

This graph shows the average maximum pupil diameter of *top curve*, the eye in total darkness and *bottom curve*, under illumination typical of an office situation. In both cases, as a person ages, his or her maximum pupil size decreases. Also shown are exit-pupil diameters (in parentheses) for several popular models of binoculars. (Adapted by author from A. L. Kornzweig, *Sight Saving Review* 24 (1954): 130–139.)

pupil of a young person's eye will expand to about 5 mm (⅕ in.) across. And on a very dark night that young person's pupil can further expand to about 8 mm (just under ⅓ in.), or slightly more.

The situation is quite different for older people. As shown in the graph, the maximum size to which a person's pupil can expand diminishes as he or she ages. This condition is known as *senile miosis*. By age forty the average maximum pupil diameter falls from 8 mm to 6 mm; by sixty it is reduced to 4 mm; and by eighty it narrows to only 2½ mm. Individuals at any age can vary by more than a millimeter from these typical values.

At low light levels, such as at twilight, the maximum diameter to which a pupil can open becomes the limiting factor that determines how well one can see. The reason is that your eye's pupil must work with the *exit pupil* of the optical in-

Objective Lens Diameter	Magnification	Exit Pupil Diameter
35 mm	7 times	5.0 mm
35 mm	9 times	3.9 mm
40 mm	8 times	5.0 mm
40 mm	10 times	4.0 mm
50 mm	7 times	7.1 mm
50 mm	10 times	5.0 mm

strument. (A binocular actually has two exit pupils, one for each of its objective lenses.) But, unlike the eye, the exit pupil of an instrument does not change its size according to the amount of available light. Its size remains fixed and is determined by dividing the diameter of the objective lens by the magnification. Here are exit-pupil diameters for some popular binocular designs; you can check them with a pocket calculator.

From this table we can see a property common to all of these instruments: They all produce exit pupils that are much larger than the 2- or 3-mm minimum diameter of the eye's pupil, and much smaller, with one exception, than the eye's 7- or 8-mm maximum diameter. Interesting, but what does it mean? Well, if the exit pupil of a binocular is larger than the pupil of the eye, some light will not be able to enter the eye. It's like trying to squirt all the water from a three-inch nozzle through a one-inch hole; some gets wasted. The image seen will continue to have the same brightness regardless of how much the size of the exit pupil exceeds the eye's pupil. However, this is true only for *extended* objects—those that have "real" dimensions: insects, houses, landscapes, the moon. It is not true for *point sources*, such as stars. The brightness of a star's image is determined by the size of the objective (front lens) of the binocular or telescope through which it is viewed. Here is a case where bigger is truly better! It's also the reason astronomers who want to see the faintest objects possible build monster telescopes having "eyes" tens of feet in diameter.

But when the exit pupil of the instrument becomes smaller than the pupil of the eye, bad things begin to happen. As the binocular's exit pupil increasingly falls short of filling the eye's pupil, the image seen grows dimmer and dimmer. This effect can be dramatically demonstrated with a spotting scope that allows you to "zoom" from low to high power. As you gradually increase the power, say from 15 to 20, the image gets bigger but maintains the same brightness. Then, as more power

Note how the exit-pupil diameter of this spotting scope shrinks as the magnification increases. This sequence was taken by changing the magnification of a zoom eyepiece from *top to bottom:* 20 to 30 and then to 45. Since the telescope had an objective lens 60 mm in diameter, 3-, 2-, and 1⅓-mm exit pupils resulted. The latter two would *not* have given a fully bright image even in the strongest illumination encountered outdoors. (Photos by author)

is added, the growing image will suddenly begin to fade. This changeover takes place when the scope's exit pupil becomes smaller than your eye's pupil.

For nature study it makes sense to have the exit pupil of your binocular a millimeter or two larger than the diameter of your eye's pupil in bright daylight. Since your subjects will usually be strongly illuminated, the "spilled" light will be of no consequence. But in deeply shaded woods or under a thick overcast, when the pupil of your eye opens up a bit, the light provided by the large exit pupil will continue to reveal details whereas a minimal exit pupil would offer only silhouettes. Finally, having an exit pupil larger than your eye's pupil is an asset at any time. The extra width means that you can see a fully illuminated field of view without having to line up the center of your eye precisely with the center of the eyepiece —an almost impossible task when you're trying to glimpse a rapidly moving object.

The *field of view* given by a binocular or spotting scope can make or break your viewing pleasure. The size of the field you see depends upon the design of the eyepiece and the *focal length* of the objective lens. (The focal length is the distance in inches or millimeters at which you have to hold the lens from a surface that acts as a screen, such as a wall, to form a sharp image of a very distant object—eyepieces aren't permitted in this experiment.)

The field of view is often cited as the number of feet a binocular can image across a target 1,000 yards away. Standard binoculars have a field of view of 350 feet or so; wide-angle ones cover about 500 feet; and super-wide-angle models expand this to 550 feet or more. Some manufacturers express, more conveniently I think, the field of view in angular measure, the number of *degrees of arc* a binocular images. A

Feet at 1,000 Yards	Degrees
300 Standard	5.7
350	6.7
400	7.6
450 Wide Angle	8.5
500	9.5
550 Super-Wide Angle	10.4
600	11.3

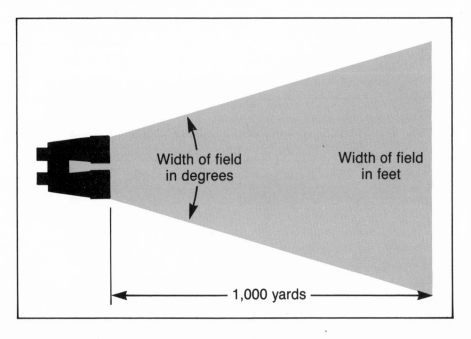

Here the concept of field of view expressed in degrees is compared to that expressed in feet at 1,000 yards.

circle has 360 degrees, so a typical binocular—one that sees seven degrees—covers about 2 percent of a circle. Another way to visualize seven degrees is to look at the full moon; since it spans one-half degree, fourteen moons can be placed side by side in a seven-degree field. As an approximate rule of thumb, to convert a field of view expressed in degrees to one expressed in feet at 1,000 yards, just multiply the number of degrees by fifty-three. The table at left matches fields of view in feet at 1,000 yards to fields in degrees.

The size of the field of view can be a double-edged sword: If you have too small a field, you will have a difficult time finding your target; if you have too large a field, the target might become lost amid the confusion of its surroundings. Nevertheless, the advantages of a wide field far outweigh its disadvantages. In addition to helping you find your subject quickly, a wide field will often contain something interesting that you hadn't noticed before—a tiny flower, an insect, a curiously patterned piece of bark. And when something moves, especially at close range, you will be able to follow it readily. A wide field is especially valuable, for example, during a heavy migration of birds when all manner of critters can be seen flitting here and there around the canopy of a tree.

It is important to ask yourself "What will be the main use of my binoculars?" If the answer is "targets at a distance," such as shorebirds on a sand spit, a narrow field of view would not be an annoyance. On the other hand, if you mainly work thickets or jungles where interesting subjects are encountered close up, a wide-angle model would be a better choice. Wide-angle viewing, especially at close range, can deliver spectacular three-dimensional panoramas. By seeing a bird or mammal in sharp focus with broad sweeps of nearby and distant brush, you are inserted into the picture and join its environment. Because larger optics are often used in *wide-field* binoculars, alternatively called *wide-angle* binoculars, these instruments are often bulkier and heavier than conventional equipment.

As we have seen, magnification plays an important role in how well binoculars function, for it directly or indirectly affects both image brightness and field of view. Thus we are faced with the question: How much magnification is the right amount? As previously mentioned, 7 power was once considered standard among bird-watchers. Today, however, they and many others would correctly argue that 8- to 10-power binoculars perform more satisfactorily for people who are able to hold such instruments steady.

This change of attitude developed rapidly in the late 1970s and early 1980s, largely through the widespread adoption of binoculars having so-called *roof prisms*, which will be discussed more thoroughly in Chapter 4. For a given objective-lens size, roof-prism instruments are significantly lighter than traditional *Porro-prism* binoculars. And since lightweight binoculars can usually be held steadier than heavy ones, the magnification can be increased without compromising stability. (Caveat: An ultralight binocular might actually jiggle about more than one having a conventional weight. Such an instrument is so light that it can't counteract normal body tremors.) Another advantage of roof-prism binoculars, in the opinions of most users, is that they can be held more comfortably, and thus more steadily, than Porro-prism designs.

Nevertheless, these advantages do not mean everyone should "power up." Consider this. If you increase the magnification from 1 (that of your unaided eye) to 7, you gain 700 percent in image scale. But if you then jump from 7 to 10, you gain only an additional 43 percent.

How close should a binocular focus? In most instances, such as viewing sporting events or spotting distant game, close-focus doesn't offer any advantage. But for others, such as bird- or butterfly-watching, the nearer a binocular can focus, the happier the user will be. In fact, under certain conditions, such as when working tropical rain forests or dense thickets, close-focus is essential much of the time.

The ability to focus really close up also means that you will get the most out of those rare occasions when you and your quarry come face to face. Only then can you glimpse the minute feather structure of a bird or nuances of color and sheen that even the best photographs might not capture. And your view will always be exceptional, for your subject will be three-dimensional and alive. Really close-up views of insects, mushrooms, mosses, and rock formations can be spectacular and exciting visual treats.

Making Optics
Work Better

The topic of binocular *coatings* is loaded with misinformation and mystery. (One widespread misconception is that coatings are put on lenses to protect them. They don't.) Yet even if you understand what a coating is intended to do, it is still possible to be misled by a less-than-forthright advertisement or sales pitch. In this section we'll see how coatings work, what types of coatings are available, and what questions you should ask to be sure you get what you pay for.

Clearly, manufacturers wouldn't cover the lenses of optical devices with some material if there wasn't a benefit. Yet, faith aside, you might wonder why a coating doesn't act like a fog and cut down the brightness of the image. In a sense it does, for no substance is perfectly transparent. However, a coating greatly offsets whatever tiny amount of light it absorbs by vastly reducing the amount of light that is reflected off your binoculars' highly polished glass surfaces. And if light rays are reflected away, they can't help form an image. Moreover, reflected light inside an optical system is doubly bad, for the scattered rays degrade an image by reducing contrast. Light is also absorbed by the glass it passes through. Little can be done about that, except to reduce the amount of glass, which will usually reduce image crispness as well. In a conventional binocular, roughly 10 percent of the incoming light is lost in passing through the prisms alone.

Binoculars have many air-to-glass surfaces where unwanted reflections take place: twelve or more are not uncommon for high-performance instruments. Typically, the objective lens has two or more surfaces, a pair of prisms four, and an eyepiece six or more.

Despite the coatings being only a few millionths of an inch thick, when you look at the surface of a lens they produce a color: purple, blue, green, or yellow. If there is *no* color, the binocular is uncoated and not worth buying. All mainstream manufacturers coat at least some of the optical elements, even on their most inexpensive models, so brand name alone will give you assurance that this important process has been carried out, at least in part.

Not all coatings are equally good, for some inhibit light reflections much better than others. No coating at all results in an approximate 5-percent light loss from each air-to-glass surface. A coating of magnesium fluoride (MgF), the most common type, reduces the loss to about 1¼ percent. Multi-coating, the most expensive and best of all processes, further reduces the loss to only ¼ percent. In the table below are

comparisons for a binocular having twelve air-to-glass surfaces. The first column shows the total light loss when no
coating is used, the second when all twelve surfaces are coated
with magnesium fluoride, and the third when all surfaces are
multicoated.

Total Light Loss after Twelve Surfaces

No Coating	MgF	Multicoated
46%	14%	3%

As you can see, an uncoated binocular suffers a 46-percent
light loss after the twelfth surface. Put another way, the brightest image an uncoated 50-mm binocular can give is only equivalent to a 35-mm binocular that is multicoated throughout. A
magnesium-fluoride coating reduces the loss to 14 percent,
which in bright light would probably go unnoticed in a head-
to-head comparison with a multicoated binocular. But in dim
light, such as when a bird or animal hides among deep shad-
ows, the extra throughput of multicoated binoculars becomes
obvious and justifies the extra cost. At such times the multicoat
acts like a race car; it always has something in reserve.

It is important to understand manufacturers' advertising lingo
regarding binocular coatings. The simple word "coated" means
very little. If only the lenses you can see are coated—the front
surface of the objective and the back surface of the eyepiece
—the binocular will resemble one that has all its surfaces
coated, but its performance will be enormously reduced compared to one that has coatings on all or most of its surfaces.

So be sure to check the manufacturer's specifications to find out how many air-to-glass surfaces are coated and with what substance. Only as a last resort, and then with caution and skepticism, should you try to elicit such information from a salesperson. Knowledgeable ones don't exist at discount stores. On the other hand, you will probably get accurate information from businesses and mail-order houses that specialize in supplying quality optical equipment to hobbyists.

To illustrate why it's important to know what goes on inside a binocular, let's look at the light loss in a "Cheap City" pair. These, we'll assume, were manufactured to counterfeit top-quality instruments by coating only the two lenses that can be seen. Notice what happens after the incoming light strikes two coated surfaces and ten uncoated ones:

Total Light Loss after Two Coated Surfaces and Ten Uncoated Ones

MgF	Multicoated
42%	40%

With magnesium fluoride the buyer would gain only 4 percent improved brightness over uncoated binoculars (46 percent minus 42 percent), and with a multicoat the gain would be only 6 percent. Nevertheless, he or she would probably have paid a premium price. The expressions "fully coated" or "100-percent coated" from a name-brand manufacturer is your best guarantee that *all* the optical surfaces have been treated with antireflection material.

Clearly, coated binoculars are vastly superior to uncoated ones. To produce a bright, vivid image under all illumination conditions normally encountered, coatings are essential. For most users, a magnesium fluoride coating on many or all surfaces will give perfectly acceptable results. Multicoated optics are for those who demand exquisite performance from their equipment and are willing to pay for it.

Fundamentals of
Binocular Design

Mechanically, binocular designs differ in one fundamental way: The two eyepieces either can be focused together or they can be focused individually. If your distance to a target changes rapidly, as in bird-watching, from very far away to close up, only simultaneous focusing should be considered. A binocular's ability to quickly deliver sharply focused images to both eyes is crucial in such circumstances. This capability is also called *center focusing*, because the images in both eyepieces are focused from the center post that holds the two halves of the binocular together.

Traditionally, the focus is changed by rotating a wheel mounted on the center post; this action moves the eyepieces in and out. There are three basic options: a narrow disk near the eyepieces, one near the objective lenses, or a one- to two-inch-long cylinder surrounding the center post. To a large measure, you only need decide which design fits your fingers best and feels the most comfortable. Yet there is one situation where the long cylinder is clearly superior: when you wear mittens or heavy gloves. Persons living in temperate latitudes or traveling to cold climes should carefully consider this easily overlooked detail. On the other hand, if you like to wear a cap with a long duckbill to shade your face or keep rain off your eyeglasses, you might prefer to have the focusing wheel near the objective end of the binoculars. In that position, when you rotate the wheel your fingers won't tend to knock the cap off your head.

How quickly should a binocular change its focus? When your subjects are apt to pop up any distance—from "infinity" down to a dozen feet or so—the best models should make the adjustment within one full rotation of the focusing wheel. Some particularly efficient designs can accommodate such a change in a half-turn or less. If you have to spin the focusing wheel through more than one rotation, the subject may have fled by the time you get a sharp image. It is also important, as you turn the wheel with your fingers, that you are able to do so without altering your grip on the binoculars themselves. If you have to adjust your hand position, you may lose your target. The disadvantage of very rapid focusing is that it may be difficult to stop turning the wheel exactly when critical image sharpness is reached.

A few manufacturers, such as Bushnell and Swift, have in some models replaced the focusing wheel with "quick-focus" devices. In Swift's design, the flip of a lever takes you from far away to close up. In the Bushnell, a rocker bar mounted

Top: With a twist of the cam-action focusing knob at center you can change the focus of the Swift 8 × 40 from close-up to far-away. (Photo courtesy Swift Instruments, Inc.) *Bottom:* The Bushnell 8 × 40 accomplishes continuous rapid focusing by means of a rocking bar. (Photo by author)

across the center post mimics the wheel's action and permits rapid focusing at any distance. Neither of these innovations has yet gained widespread acceptance among naturalists, though there seems to be nothing inherent in such designs that would preclude their usefulness in the field. In fact, a rocker-bar assembly offers a real advantage when you wear mittens or gloves.

Most roof-prism binoculars, which are described below, employ a scheme whereby internal, unseen moving lenses do the focusing. As a result, the eyepieces remain in a fixed position; they don't move in and out. Such a design, with its fewer openings to the outside world, means that roof-prism binoculars are inherently more moisture-resistant and airtight than are Porro-prism models. Nevertheless, suction caused by the moving lenses can draw humid air into the body of the binocular and thereby fog the optics from the *inside*. This is the worst possible condition, for drying out could take hours or even days!

Three basic designs for binocular optics have evolved: Galilean (named after the famous Renaissance Italian physicist Galileo Galilei), Porro-prism (named after another, but very obscure, Italian scientist; his first name might have been Ignatz but no one is sure), and roof-prism (named after the shape of the prism). This order is also roughly that of the binoculars' price tags.

Galilean models, often referred to as "opera glasses," are the cheapest and least satisfactory binocular under any circumstance, outdoors or in. They can only function effectively at very low magnification, 4 times or less, and have a very constricted field of view. This kind of binocular does its work with lenses alone—there are no prisms—so its barrels are straight. In this regard, Galilean binoculars look superficially like the expensive roof-prism models.

When you picture a binocular in your mind's eye, the image you conjure up will probably be that of a Porro-prism design. These binoculars, traditionally held by characters in Hollywood war and racing movies, have been the mainstay of manufacturers for decades. And they will almost certainly continue

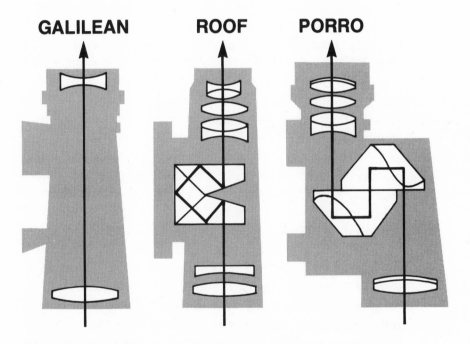

GALILEAN   **ROOF**   **PORRO**

The light path through the three basic binocular designs. *Left:* The primitive Galilean type with its straight-through optical path. *Middle:* The popular contemporary roof-prism design has a shape that resembles the Galilean. *Right:* A Porro-prism binocular is relatively bulky but usually delivers the sharpest images. (Diagram by author)

to dominate the marketplace in the foreseeable future. There are two reasons why this is so. The first is their ability to consistently achieve both a wide field and excellent image sharpness. The second is cost: high-quality Porro-prism binoculars can usually be purchased for less than half the price of equivalent models that use roof prisms.

But roof-prism binoculars have many virtues of their own. Four mechanical features stand out: they are generally lighter than their Porro-prism counterparts; they tolerate rough treatment better because the optics are arranged more simply; they are less susceptible to internal fogging due to their greater structural integrity and method of focusing—in other words, they don't have a "bridge"; they have a shape that makes them almost beg to nestle in your hands. In fact, many people claim that they don't sense holding roof-prism binoculars. Perhaps that impression is gained from the relative lightness of this design and its straight, hand-fitting barrels.

As mentioned earlier, a disadvantage of roof-prism binoculars is the high cost of the best models. And there is another, even more important, "kicker." The images formed by roof-prism instruments, in general, will not be as sharp as those produced by Porro-prism designs. To achieve sharpness equal to Porros, roof prisms must be manufactured and aligned about three hundred times more accurately. On a commercial assembly line that tolerance cannot be routinely achieved. Another problem inherent in roof prisms is that they have a special reflecting surface that can cause a 5- to 15-percent light loss above and beyond that of any other optics in the system. The result, of course, is a dimmer image. Most manufacturers use aluminum for this surface; a silver coating, such as Bausch & Lomb employs in its Elite product line, is superior.

A problem with one of the most popular roof-prism bin-

A popular roof-prism binocular is this 10 × 40 made by the West German firm of E. Leitz. This unit has its eyecups extended, as they would be for a non-eyeglass wearer. In some roof-prism designs, such as this one, dioptre corrections are made by rotating a wheel on the lower end of the center post rather than at the eyepiece. (Photo courtsey E. Leitz, Inc.)

oculars, the 10 × 40 by Leitz, is that standard off-the-shelf units cannot focus closer than about thirty feet. To get the close-focus down to twenty-two feet requires a relatively expensive retrofit, one that may take your binoculars away for weeks. In my opinion, a close-focus of twenty or thirty feet is totally unacceptable if you want to view nature's subtle details. Many excellent binoculars, particularly so-called compact models, focus down to fifteen feet and some can get well under ten feet.

Before leaving the topic of binocular design, I want to point out that not all prism glass is created equal. The least satisfactory material is called *ordinary crown*, which produces a dimming at the edge of the field of view. It is easy to determine whether this type of glass has been used. Simply hold the binoculars a foot or so in front of your eyes and look at the exit pupils. If they have grayish, angular edges—more or less turning the circular exit pupil into a square—then the prisms are of ordinary crown. Better-quality and more expensive glasses that deliver a uniformly bright field are called *flint* or *barium crown*. The latter (often abbreviated BaK or BAK) is used in top-notch binoculars. In particular, BAK-4 glass is touted by many manufacturers as the best available.

Binoculars for
Everyone

I've already made it clear that no one binocular is the best choice for everybody. It would be easy to list the specifications of scores of binoculars that perform well, but such a mass of statistics would tend to confuse rather than illuminate. So, in the table at the end of this chapter I've limited the selection to a relatively few models, most of which I have tested and found to give exceptional performance or to be a particularly good value. My choice is a very personal one—there are many other binoculars that perform exquisitely. You should consider this list as representative of the best that's available, rather than as definitive.

I was aided in making this selection by a survey conducted

by *British Birds* magazine (April, 1983); a series of critical tests performed by members of the New Jersey Audubon Society (results were published in the *Peregrine Observer*, that organization's newsletter, in Fall 1984); and a Consumer's Union appraisal in *Consumer Reports*, July, 1989.

The binoculars in the table are appropriate for a wide range of outdoor activity, and many have garnered a broad following by proving themselves in the field time and again. The models listed can act as reference standards against which you can compare the performance of other instruments. To make the opportunity for intercomparison as broad as possible, I've included at least one product from most major manufacturers. Before you buy any optical instrument, be sure to check the manufacturer's current guarantee/warranty policy and understand what repair services are available. (Addresses and telephone numbers of many manufacturers are given in Appendix A; I recommend you obtain their catalogues.)

All the listed binoculars have fundamental technical specifications that guarantee excellence in the field. Except for two giant binoculars suggested for astronomical observation and two compact binoculars, all are 7 to 10 power and give exit pupils (EP) greater than 3½ mm in diameter. The so-called *twilight factor* (TF) indicates how well a particular binocular will perform under low-light levels, such as at dusk. It is calculated by taking the square root of the quantity (magnification × objective diameter in millimeters). The larger this number, the brighter will be the image under poor lighting conditions. However, be careful how you interpret the twilight factor: A high-quality coating will improve a particular model's twilight performance relative to an identical binocular with a coating of lesser quality.

ER is the eye relief in millimeters as given by the manufacturers; values smaller than 17 or 18 will usually not give a full field of view for eyeglass wearers. The close-focus (CLF) distances for binoculars recommended for bird-watching are from independent tests; the others are from information supplied by the manufacturers. No close-focus data are given for binoculars recommended for astronomy or marine use, since this information is largely irrelevant.

The field of view (FOV) cited is that given by the manufacturers, and it applies to persons who do not wear eyeglasses. After testing scores of binoculars, the New Jersey group and I both found that the field actually seen is usually smaller than that claimed by the manufacturer—about 10 percent smaller. Therefore, before buying any binocular, be sure it gives a satisfactory field of view; this is most easily done by comparing several models. Also, be aware that certain binocular designs penalize eyeglass wearers much more than do other designs (see Chapter 7). So, if you wear eyeglasses, check the field size with them on and off; you might be surprised at the difference. The good news is that manufacturers are offering more and more models especially designed for eyeglass wearers. Be sure to try out these so-called long-eye-relief binoculars.

The weights included in the table span a wide range, even if you ignore the sixty-pound monster from Fujinon. Compact binoculars are truly feather-light—you don't even sense 9 to 11 oz hanging around your neck. Binoculars recommended for general nature study and bird-watching bunch quite tightly, from 21 to 29 oz, and can be carried all day without strain even if supported only by a narrow strap. Heavier instruments, say from 30 oz and up, require wide straps for proper support and even then they could be a real pain in the neck, literally, on a long field trip. Most binoculars that are particularly effective for astronomy and marine use are very heavy because of their large size or rubber armor. At best they can be hand-held for only very short periods, and usually they require some kind of mechanical support.

The next-to-last column lists the manufacturers' suggested list price for 1989. Of course, these prices can change at any time and are usually roughly twice that charged by many reputable retail outlets and mail-order houses. The final column indicates whether a particular binocular has Porro prisms (PP) or roof prisms (RP), center focus (CF) or individual focus (IF), and whether it is armored (A).

The following sections explain some of the criteria I used in making my selections. There are also some tips on what you should look for when choosing a binocular for some specific purpose.

Astronomy = Big Binoculars

Most of us live in urban settings with skies polluted by smoke and outdoor lighting. Nights never become really dark. Yet even under the horrid conditions that prevail in metropolitan areas you can see bright stars and planets and our moon. (Of course, during the daytime you can see the sun, too, but you should *never* look at it directly with your unaided eye or with any optical equipment; severe eye damage or blindness could result.)

Any binocular or spotting scope can be used for casual sky-watching, and many will give surprisingly good views. Even your 7-power binoculars, if they are held steady, will show the rings of Saturn. But if you get far away from city lights and experience a truly dark sky, you might want an instrument particularly suited to astronomical observing. Probably not one city dweller in a thousand has experienced a sky so black that the combined light from the stars themselves brightens the landscape. If you find yourself under such a sky, don't go to bed but take time to look; you'll stumble across all kinds of fascinating sights. A fuzzy glow might unfold into a swarm of stars; even grander, the gauzelike glow of the Milky Way will dissolve into myriad stellar lights.

As mentioned earlier, bigger is better for astronomy, and in the table I've included two large binoculars and one giant one that produce particularly fine celestial images. In addition to their large size, astronomical optics can often profit from much higher magnifications than can other instruments used outdoors. High power can often help make faint objects visible, and sometimes high power is absolutely necessary in order for you to see the tiny disks of planets or to separate closely adjacent stars. Of course, such high-power instruments need to be mounted on a tripod or other firm support. But remember, you don't *need* special equipment to enjoy the heavens.

Binoculars for Bird-Watching

In field tests, all of the instruments listed in this section demonstrated extremely high resolution, better than 0.6 minute of arc (.003 percent of a circle's circumference or ⅟₅₀ of the full moon's diameter). Theoretically, the objective lenses of these binoculars are capable of ten times better resolution, but their magnification is much too low to test that limit. The ability of a binocular to focus on nearby objects is also extremely important to bird-watchers; thus the minimum close-focus (CLF) distance in this section was determined by direct measurement in the field. Hence the values cited here might differ from the manufacturer's specifications.

These 8 × 23 Nikon compact binoculars weigh less than a pound. They stand as tall as a pack of 100's cigarettes and are twice as wide. (Photo courtesy Nikon Inc.)

Compact Binoculars Are Not Toys

The main virtue of these diminutive instruments—often they are not much larger than a pack of cigarettes—is that they can be carried anywhere, anytime. And there is another important plus: Many compacts can focus down to only a half-dozen feet or so, which permits truly "microscopic" views of insects and other tiny things that you only notice when you're right on top of them. A pair of compacts should be in every naturalist's pocket or purse, where they can instantly be whipped out to take advantage of accidental viewing opportunities, such as when commuting, changing classes, or shopping. Compacts are also excellent for watching sporting and theatrical events —vastly better, in fact, than the traditional Galilean ''opera glasses,'' which are about the same size.

Compact binoculars should not be dismissed just because they are small and look ineffective. In particular, you should consider models with powers between 6 and 8 and apertures between 20 and 30 mm. Many have magnifications equal to larger optics, and their exit pupils (roughly 3 mm in diameter) are perfectly adequate for bright-light conditions. A small point, but one that might influence your decision as to which compact to buy, is that some designs have the objective lenses closer together than the eyepieces. In such cases the stereo or 3-D effect of binocular viewing is compromised. A scene with subjects both near and far will appear flat and will lack depth, looking rather like a painting from ancient Egypt.

Having made a case for compacts, I want to stress that your first choice in binoculars should always be a standard model, such as one of those included in the table under bird-watching or general nature study. Compacts are special-purpose instruments; they don't replace conventional equipment!

Nature and Everyone's Binoculars

The three instruments listed in this section of the table have no special features—except quality. Overall, their properties are ideal for viewing everything in general and nothing in particular. These are truly all-purpose instruments that won't disappoint. They are also good choices for those who are uncertain about what they want or how their new binoculars will be used.

Waterproof Binoculars

Under certain circumstances, such as when you are aboard a spray-drenched boat or in a steamy jungle, a truly waterproof binocular may be essential. Throughout my earlier discussions of the most desirable binocular features, I have stressed the advantages of center-focus systems—ones that focus both eyepieces simultaneously. But when waterproofing is your first priority, it might be necessary to choose a binocular with individual-focus eyepieces. Center-focus binoculars generally have too many places where moisture can seep or be drawn into the optical assembly, though some of these designs are advertised as waterproof or water resistant.

Several manufacturers offer truly waterproof binoculars, whereby the optics and moving mechanical elements are sealed with O-rings. Often these models are purged with inert nitrogen gas to prevent internal fogging from sudden changes of temperature, humidity, or both. Of course, such specially constructed binoculars will also perform excellently in dusty or other unusually harsh environments.

How troublesome can moisture be? Once, on Saint Lawrence Island in the middle of the Bering Sea, my equipment became drenched whenever I entered a warm shelter; it was as though I had left my binoculars, spotting scope, and camera out in a rainstorm. On the very first day my friend's rather old roof-prism binoculars fogged internally and stayed that way for the remainder of our visit. Fortunately, I had taken a pair of compact binoculars along as a backup, so I was able to get him out of a real jam.

One manufacturer, Steiner of West Germany, specializes in producing waterproof binoculars for rugged military use; some of these even come with a built-in compass. In fact, its light-

These Steiner 8 × 30G binoculars are armored and waterproof, with individual-focusing eyepieces. (Photo courtesy Pioneer Marketing & Research, Inc.)

weight "fiber-reinforced polycarbonate" body is described as unbreakable. Although this firm is the largest binocular manufacturer in Europe, it is relatively little known in the United States. Certainly, its specialized optics would not be chosen for everyday use, but they could prove invaluable for persons anticipating extreme environmental conditions. Steiner's 8 × 30G model is particularly attractive due to its 3.8-mm exit pupil, 6.8-degree nominal field of view, and extreme lightness (only 17 oz, including armoring).

Individual-focus binoculars, particularly 7 × 50 models, are standard equipment aboard pleasure boats and are also widely used by amateur astronomers. This same configuration might also be preferred for some specialized types of nature study, such as watching birds that never come close up. Hawkmigration studies from a mountaintop is one obvious application; another would be counting seabirds or waterfowl. Individual-focus binoculars work equally well when the subject always stays at the same distance, such as when you want to monitor a bird's nest from your living room. As long as your subjects don't come too close—less than twenty or thirty feet—some individual-focus binoculars have more general uses; their optical systems have been designed to produce sharply focused images all the way from that minimum distance to infinity. In other words, one focus-setting does it all.

Zoom Binoculars

These curious devices can change magnification from about 7 to 15, but they have not gained a wide following. The reason is clear. For the privilege of zooming, fundamental qualities of binocular design have had to be compromised, such as weight and field of view. At their best, binoculars should be lightweight and wide-angle—to diminish these features is to reduce overall usefulness. I have never used a zoom binocular that I liked, so none have been included in the table.

Table A: **Some Exceptional Binoculars for Nature Study**

Mfg.	Model	EP (mm)	TF	ER (mm)	CLF (ft)	FOV (dg)	WT (oz)	$	Notes
Astronomy									
Nikon *Prostar*	7 × 50	7.0	18.7	15		7.3	52	1,058	PP,IF
Nikon *Astrolux*	10 × 70	7.0	26.5	15		5.1	88	1,015	PP,IF
Steiner	15 × 80	5.3	34.6	11		3.7	56	1,129	PP,IF,A
Fujinon	25 × 150	6.0	61.2	17		2.7	970	11,000	PP,IF
Bird-Watching									
B&L *Elite*	8 × 42 +	5.2	18.3	20	12	7.0	27	1,479	RP,CF
Mirador	8 × 42ZCF	5.2	18.3	18	15	6.6	22	261	PP,CF
B&L *Discoverer*	9 × 35	3.9	17.7	14	9	7.3	23	590	PP,CF
Zeiss	10 × 40B +	4.0	20.0	18	16	6.3	25	1,085	RP,CF
Compact Binoculars									
B&L *Custom*	7 × 26	3.7	13.5	17	12	7.3	11	388	PP,CF
Nikon *Venturer II*	8 × 23	2.9	13.6	10	16	6.3	9	132	PP,CF
Minolta	9 × 24	2.7	14.7	—	—	5.7	11	166	PP,CF
General Nature Study									
Zeiss	7 × 42B/GA	6.0	17.1	18	12	8.5	28	1,030	RP,CF,A
Swift *Audubon*	8.5 × 44	5.2	19.3	15	15	8.2	29	375	PP,CF
Leitz *Trinovid*	10 × 40B +	4.0	20.0	13	22*	6.3	21	1,335	RP,CF

Table A: Some Exceptional Binoculars for Nature Study

Mfg.	Model	EP (mm)	TF	ER (mm)	CLF (ft)	FOV (dg)	WT (oz)	$	Notes
Marine									
Swarovski *SLC*	7×30B	4.3	14.5	18		7.4	19	530	RP,CF
Fujinon *MTR-SX*	7×50	7.0	18.7	17		7.5	49	600	PP,IF,A
Steiner	8×30G	3.8	15.5	11		6.8	17	265	PP,IF,A

EP   Exit pupil diameter
TF   Twilight factor (see text)
ER   Eye relief (see text)
CLF Close focus distance
FOV Field of view diameter
WT Weight
$ Manufacturers' suggested list price, 1989
PP   Porro-prism design
RP   Roof-prism design
IF   Individual-focus eyepieces
CF   Center-focus eyepieces
A   Rubber armoring

*After close-focus retrofit
+Armored model also available

Using Your
Binoculars

All quality binoculars have pretty much the same ancillary features, those little things that make them work better or more conveniently. You should take care to understand what these features do, so that you can get the best performance out of your equipment.

Interpupillary Scale

When you look at the eyepiece end of a pair of binoculars, you will usually see a small disk atop the center post marked with numbers between about sixty and seventy. These numbers indicate the distance, in millimeters, from the center of one eyepiece to the center of the other. As you swivel the binoculars to spread them apart or close them down, an index mark on the body of the binocular tells what the setting is. (Caution: Not all manufacturers carefully align these scales. Even if they were set correctly at the factory, with normal use they may slip out of adjustment.)

This interpupillary scale enables the binocular-user to match the centers of the eyepieces to the centers of his or her eyes —not everyone has the same head shape! In fact, a friend of mine has eyes set so close together (only 54 mm) that he cannot get the images of some binoculars to blend together. So it is crucial when buying binoculars to make sure the separation of the eyepieces can be made to match that of your eyes.

The dioptre scales on these 7 × 50 Swift Mariner compact binoculars are boldly marked on both of the adjustable eyepieces. By turning to a negative (−) setting, you can correct for nearsightedness; a positive (+) setting will correct for farsightedness. Notice also the well-marked interpupillary scale at the top of the center post. (Photo courtesy Swift Instruments, Inc.)

If the eyepieces are too close together, the image you see will appear dimmer than it should be; also, the field of view will be restricted. Too wide a separation will cause a dark line to pinch the middle of the field of view or even divide it into two parts. Such misadjustments ruin the performance of your binoculars and cause eyestrain that can lead to headaches. In my experience, roof-prism designs seem to require more careful interpupillary settings than do binoculars with Porro prisms.

An optometrist measures your interpupillary distance during a routine eyeglass fitting. If you have an old prescription, the interpupillary distance will be written on it; otherwise, call your optometrist, for he or she should have this information on file. You can also measure your own interpupillary distance by placing a millimeter ruler across the bridge of your nose and having someone read off your eyeballs' center-to-center separation.

Dioptre Scale

Most people's eyes do not focus exactly the same, even if corrective eyeglasses are not required. In everyday life subtle differences between two individual eyes' nearsightedness or farsightedness go unnoticed. But when a binocular magnifies an image, even a small mismatch becomes conspicuous.

For this reason binoculars have adjustable eyepieces, one for center-focus designs and two for individual-focus models. Calibrations typically range from $+2$ to -2 and they count *dioptres*, the unit of measurement an optometrist uses to prescribe the amount of correction for eyeglasses. Dioptres measure how much the effective focal length of an optical system is lengthened or shortened. So by turning the adjustable eyepiece you can correct for any slight nearsighted or farsighted imbalance between one eye and the other. In effect, you change your eye's focal length.

However, you have to be careful when you attempt to find the right dioptre setting. Here's the problem. When you look through a binocular and the image from one of its halves is out of focus, your eye will try to compensate by refocusing itself. If the misfocus is only slight, the eye will actually succeed in creating a sharp image. But an unwanted by-product will be eyestrain, which will recur every time the mismatched binoculars are used.

It's amazing that many owners of binoculars remain unaware that one simple adjustment can take the fuzziness out of an image. Although it may seem a bit complicated, I'm going to describe a technique that will make sure both of your eyes are focused perfectly. Take a pair of center-focus binoculars and follow me step by step. (For individual-focus models you will have to adjust each eyepiece separately, but follow the same procedures as described.)

First, pick a target at least several hundred feet away that has sharply defined features, such as a billboard or a chain-link fence. Then use a lens cap or the palm of your hand to cover the front lens that "feeds" the adjustable eyepiece. Now, *quickly* focus the other lens, using the center-post mechanism. (If you focus the binocular too slowly, your eye might try to focus itself and thus thwart your efforts.) You must also keep your viewing eye relaxed, as described in the next paragraph, or, again, it will try to focus itself. Finally, you must not squint with either eye or shut the one that's not being used.

After getting the viewing eye into sharp focus, it's a good idea to cast both eyes at the ground for a few seconds to relax them. Now look through the nonadjustable eyepiece again. Is the image still sharp? If so, good! If not, repeat the entire

How to get both eyes into sharp focus, as demonstrated by the author's wife, Caroline. 1. Cover the objective lens that feeds light to the adjustable eyepiece (the one with the dioptre scale), and then focus on a distant target.

2. Rest both eyes by glancing away for a few seconds.

3. Without touching the focusing knob, look at the same target but this time with the objective for the nonadjustable eyepiece covered. Turn the *adjustable* eyepiece until sharp focus is attained, and then rest your eyes again. Now, when you look through both eyepieces simultaneously, the images from the two halves of the binocular should be sharp. If they are not, or if one eye "feels uncomfortable," repeat the procedure. (Photos by author)

focusing process. After you are certain that the image in the nonadjustable eyepiece is really sharp, take a break for a few seconds and again relax your eyes by glancing at the ground. Now you're ready for the next step.

Without moving the center-post focusing mechanism, cover the objective lens of the nonadjustable eyepiece. Then look through the adjustable eyepiece and turn its barrel, first one way and then the other, until you get a sharp focus. As before, keep both eyes open and relaxed during this procedure. Try focusing several times, creeping up on perfection as it were, until the image remains sharp whenever you look into the adjustable eyepiece. (Zeiss employs a second wheel on the center post to balance the focusing of the two halves of its popular 10 × 40 roof-prism binocular. Nevertheless, the principles for precise adjustment are the same as described above.)

If you did everything correctly, when you look through both eyepieces simultaneously the image should be in perfect focus. (A *tiny* adjustment using the center-post focusing mechanism is okay at this point to fine tune the system.) Your eyes should feel entirely comfortable. I cannot overstress the importance of getting your binoculars into perfect adjustment to enhance your viewing pleasure and inhibit eyestrain.

The adjustable eyepiece on most binoculars has a screw thread. This design has the advantage of permitting any setting within the dioptre range provided. Its disadvantage is that the setting can become easily knocked out of whack, such as by accidentally brushing the eyepiece collar with your hand. (This happens frequently when a binocular is put into or taken out of its case.) So, once the proper dioptre setting has been found, check it every time you use your binoculars. A better idea, one that I've used for years, is to secure the adjustable eyepiece with a piece of smooth electrical tape; it's waterproof and extremely durable.

A relatively recent and, in my opinion, superior design for the adjustable eyepiece features click-stops. Although you can only choose dioptre settings at specific intervals, the increments are so small that there is no problem getting a perfect match to your eyes. This innovation makes it difficult to inadvertently throw your adjustable eyepiece out of focus.

Tripod Mounts

Many high-power or heavyweight binoculars have a built-in fitting (¼ in. in diameter, 20 threads to the inch for the United States market) that enables you to mount the instrument on a

The fixture for attaching a binocular to a tripod should be sturdy and adequate for the weight of the instrument, such as on this Swift 8.5 × 44 Audubon. (Photo by author)

tripod. Although of little or no use to most people, this fitting will serve you well if you enjoy such activities as target shooting or astronomical observing. If you think you might benefit from a tripod mount, make sure it is rugged and solidly integrated into the binocular body so that it will not become easily damaged or tear out.

Under normal circumstances, people with reasonably steady hands and equipment appropriate to their body frames should be able to readily manage 9- or 10-power binoculars. A well-chosen binocular shouldn't need a crutch! Having said that, here are a few tips on how to hand-hold a binocular properly. If you don't do it correctly, you won't get the best views— period. For example, when I see someone holding a pair of binoculars with only one hand, I know that he or she is either a beginner or an awfully slow learner. Experienced users know that their aim is most accurate and the view most steady when both hands are put to work, one on each barrel.

That reasoning is pretty obvious when you think about it, but the mechanics of using binoculars properly go even further. Consider, for example, how your body and binoculars work together. If you hold the binoculars with your arms thrust out sideways, on a horizontal line with your nose, your arms will tire quickly and soon your hands will begin to tremble involuntarily. A much better technique is to let your arms fall naturally into a more or less vertical position, roughly parallel to your body. Gravity is now working for you rather than against you. Your arms tend to become pressed toward your body and, with your hands, they form a triangle that stabilizes the binoculars. This simple mechanical action is one reason why the roof-prism designs have gained such favor: their cylindrical shape almost forces you to choose the proper hand position. Finally, don't use a fingertip grip. Hold the binoculars

Two ways *not* to hold binoculars. *Top:* An inadequate fingertip grip is being used on an old Swift 8 × 40 Saratoga. *Bottom:* The Bausch & Lomb 8 × 42 Elite is being gripped correctly, but the arms are thrust too far forward for maximum stability. (Photos by author)

firmly and wrap your fingers around the barrels. But be careful not to tighten up too much; if you become too aggressive for a long period of time, fatigue and tremor will result.

For Eyeglass
Wearers

Eyeglasses present a special problem to binocular users, especially if they are needed to correct for nearsightedness. (For reasons that will become obvious shortly, none of my comments directed to eyeglass wearers applies to people using contact lenses.) The basic problem is that most nearsighted persons have to keep their glasses on in order to see well enough to place their subject in a binocular's field of view. If they take their eyeglasses off, the scene becomes so blurred that the subject is lost. Nevertheless, I must admit that some people have developed the knack of raising their glasses to their foreheads with their fingertips while simultaneously lifting their binoculars to their eyes. (A report in the April, 1983 issue of *British Birds* magazine indicates that about 40 percent of its readers who wear eyeglasses manage this feat.) Yet this routine has always seemed awkward to me, perhaps because it demands more dexterity than I possess. But no matter, for the entire issue of whether or not eyeglasses should be lifted can be easily avoided by choosing your binoculars wisely.

It is important to understand that with any binocular you will be able to see the full field of view only if your eye is positioned at one special distance behind the eyepiece. If you're curious, you can locate this place, called the *eyepoint*, by

With a spotting scope pointed at the sky, a piece of waxed paper was slowly moved backward from the rear edge of the eyepiece. This picture was taken when the spot of light formed on the waxed paper was sharpest and reached its smallest diameter. This distance, between the paper and the rear lens of the eyepiece, marks the eyepoint of the optical system. (Photo by author)

aiming a binocular at a brightly lit sky while moving a piece of waxed paper toward and away from the eyepiece. A spot of light will appear on the waxed paper; when its diameter is smallest and the image sharpest and brightest, you've found the eyepoint. The distance between the eyepiece lens and the sharp image on the waxed paper is technically known as the *eye relief* of the system. In general, the greater the eye relief, the better off you'll be.

In order to get a full field of view with ordinary eyeglasses, the eye relief of binoculars should be at least 17 mm. Persons with especially thick glasses might need even greater eye relief, up to 21 mm. Here is where a problem arises, for many standard binoculars have only 12 to 14 mm of eye relief, and wide-angle models can be even worse with only 9 or 10 mm.

For some inexplicable reason, almost all manufacturers have traditionally avoided citing in their ads or brochures the amount of eye relief a binocular has, which would allow eyeglass wearers to make much more informed purchases. European manufacturers such as Leitz and Zeiss have given some help by indicating with a "B" in their product codes binoculars that are especially designed for eyeglass wearers. And just recently several other firms have begun to market long-eye-relief models.

Since it's awkward to measure the eye relief in a department store (who carries waxed paper and a millimeter ruler around with them?), you'll have to improvise. Here's an easy way to determine if a particular binocular will work well for you. If your eye is positioned inside the eyepoint, the image you see will be dramatically darkened. But when your eye is outside —as it often is for eyeglass wearers—the field of view will become constricted, sometimes considerably so. Thus, a person without the need for corrective lenses might be able to see wide vistas through a certain binocular, but someone wearing eyeglasses might get only a keyhole perspective.

There is no simple rule to guide eyeglass wearers to the correct binocular design. The degree to which the field will constrict when eyeglasses are used depends partly on the manufacturer's efforts to accommodate such persons—that is, by providing extra eye relief or especially shallow eyecups—and partly on the very design of the binocular itself. Although modern binoculars are intended to be used both with and without eyeglasses, not all designs do the job equally well. It's a game of trade-offs and costs, so try a wide variety of models to find ones well suited to you.

Two methods are used to satisfy everybody's eyepoint-positioning needs: solid, rotating eyepiece caps that screw in and out, or flexible-rubber shields that can be folded down. Both designs are intended to keep the eye of the user at the critical eyepoint whether eyeglasses are worn or not. For non-eyeglass wearers, the caps or shields should be fully extended. Eyeglass users should compress these devices to compensate for the distance the glasses separate the eye from the eyepiece.

Flexible shields are very common today and are greatly preferred by eyeglass wearers, for they tend to scratch much less than the old-fashioned hard caps. Nevertheless, even rubber shields can nick your glasses (especially the lightweight plastic variety) because they still pick up bits of dirt. No matter which alternative you choose, if you use binoculars frequently, after only a few months you'll begin to notice frosty rings forming on your glasses. The real disadvantage of rubber shields is that they begin to deteriorate after only a couple of years' use or even less, and replacements for some brands are quite expensive.

Some binocular designs accommodate all users—the field

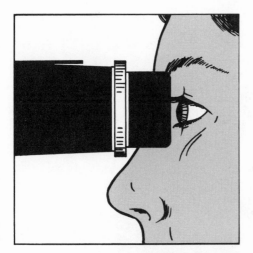

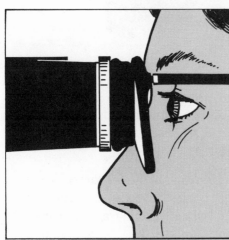

Top: Someone who does not wear eyeglasses should use his or her binoculars with the eyepiece cups fully extended. *Bottom:* To preserve the greatest possible field of view, eyeglass wearers should fold the cups down.

of view seen when wearing eyeglasses is the same, or very nearly so, as without them. But the majority of designs restrict the field significantly when eyeglasses are worn. I imagine that many people who wear eyeglasses miss a lot of exciting

The Bausch & Lomb 9 × 35 Discoverer has proved to be one of the most successful Porro-prism models ever made. Its optics are noted for their extreme sharpness. Unfortunately for persons who wear eyeglasses, this binocular's short eye relief results in a constricted field of view. (Photo courtesy Bausch & Lomb)

moments because they have "binocular tunnel vision" and don't even know it. One look is all it takes to appreciate a truly wide field of view. Thus, when selecting a new pair of binoculars, an eyeglass wearer should try out *both* standard and wide-angle designs. The standard models have an inherently smaller field of view, but their greater eye relief might offset this limitation. On the other hand, wide-angle and super-wide-angle binoculars might win out despite their inherently shorter eye relief. What you get depends on the total design of a binocular and not on any one property.

You should evaluate super-wide-angle binoculars (with fields of ten degrees or more) very carefully and exactly as they will be used in the field; that is, with your eyeglasses on. To non-eyeglass wearers, these super-wide-angle models may perform poorly, giving curved, fuzzy, or colored images at the edge of the field of view. But these faults should not concern most eyeglass wearers, for they will not see the field's edge because of restricted viewing due to the short eye relief of these optical systems.

The latest entry into the "super binocular" category is the Bausch & Lomb 10 × 42 Elite. In addition to an adequate-size exit pupil (4.2 mm) and field of view (5.6 degrees), it has excellent eye relief for eyeglass wearers and focuses very quickly from "infinity" to close up. (Photo courtesy Bausch & Lomb)

Spotting Scopes

Next to binoculars, a spotting scope is the best friend of anyone who must view very distant objects or get really close-up views of nearer ones—such as when scanning for big game or watching shorebirds or hawks. In fact, under such circumstances a spotting scope is essential. The reason, of course, is that it provides the high magnification necessary to find distant subjects or to distinguish one species from another.

The great value of a spotting scope has been known to the military and to target shooters for many decades. They were first introduced to American bird-watchers in the 1950s, and today all serious aficionados have one. Curiously, these instruments took much longer to catch on in Britain and Europe.

Yet, a survey conducted in 1983 by *British Birds* magazine showed that over 80 percent of its readers now use a spotting scope.

But beyond those instances when distance makes a spotting scope an absolute necessity are the occasions when it can give you magnificent "microscopic" views of your subject. When such instruments are steadied by a tripod, car-window bracket, or even a hand-held gunstock mount or shoulder brace, they can reveal details no binocular can match. For many bird-watchers, such views are becoming almost mandatory as field identification becomes ever more sophisticated. A mounted spotting scope can also allow you to watch, tirelessly and for hours on end, a special activity such as nest building.

The Kowa TSN-4 77-mm spotting scope features fluorite lens elements and is becoming popular among expert bird-watchers. Here it is set up for photography; the tubular adapter creates an 800-mm f/10 system. A sister scope, the TSN-3, has the eyepiece offset at a forty-five-degree angle from the optical axis; for convenience at all viewing positions it can be rotated. (Photo courtesy Kowa Optimed)

The selection of a spotting scope for bird-watching, target-shooting, or viewing a yacht race is rather easier than choosing binoculars. For one thing, there is less real variety among models (see Appendix A for addresses of most major manufacturers). But unlike binoculars, which use only lenses to form the image, spotting scopes sometimes combine lenses and mirrors. Either system can perform superbly.

Lens-Type Spotting Scopes

Everything said about binoculars—such as image sharpness, lens coatings, a wide and undistorted field of view—applies to lens-type scopes. Optically, you should judge them just as you would a binocular. There is one recent wrinkle, however, and it pertains to the type of glass used for spotting-scope objectives. Connoisseurs are now paying a lot of attention to so-called fluorite lenses, which have elements made from crystalline calcium fluoride. One of the major producers of these lenses is Kowa, which claims "images that are sharper and brighter, with unmatched image clarity and color fidelity." There is no doubt that fluorite can deliver superior performance, such as almost complete freedom from spurious color, but so can other glasses. So the question is: Do the gains from fluorite offset its extra cost, which is two or three times that of conventional optics? I don't know the answer. Like so many other facets of optical performance, it's a matter of personal preference. However, in a head-to-head comparison with Kowa's TSN-4 fluorite-lensed spotting scope, my old Bushnell Spacemaster did just fine. One thing does seem certain, however: Market forces and the bandwagon effect are sure to bring us ever more fluorite telescopes in the future.

The Astro-Physics StarFire 5-in. f/8 refractor incorporates state-of-the-art optical design for astronomical observing and photography. (Photo courtesy Astro-Physics)

Mirror-Lens Scopes

Since the 1960s, mirror-lenses (also called *catadioptric systems*) have gained enormous popularity. The application of this design ranges from giant astronomical telescopes with apertures of 14 in. or more, to compact long-focus lenses for 35-mm cameras. This design has some advantages over traditional optical systems that use only lenses. For example, when a lens and mirror are combined, the length of the spotting scope's tube can be made much shorter relative to an equivalent instrument that uses only conventional lenses. One potential drawback, though not of great concern in well-made units, is that a hard knock can cause mirror-lenses to go out of adjustment more easily than such a blow would affect a pure-lens system.

Provided the instrument is of high quality, it doesn't much matter which type of spotting scope you choose. However, because mirror-lens systems are often designed for amateur astronomers who are comfortable looking at the sky upside down or right for left, be sure any model selected for terrestrial viewing is capable of producing an image that is right-side-up and, of equal importance to many, has the correct right-left orientation. A world-famous maverick that fails the last test is the Questar, which, not surprisingly, was originally designed for amateur astronomers. Nevertheless, it remains a favorite instrument among professional tour leaders and others who can afford such expensive equipment; mechanically and optically, it is one of the finest commercial scopes ever made. (The recently introduced Kowa TSN-4 77-mm spotting scope is now giving Questar a run for its money among natural-history-tour leaders.) Many telescopes designed for astronomy work well for terrestrial viewing. But all of them require an *erector*, an optical device that delivers a right-side-up image.

The Questar 3½-inch-aperture telescope is regarded by many as the Cadillac of spotting scopes. Note the deeply curved front lens with a mirrored spot at its center, the so-called Maksutov optical design. This model is set up for photography, and it also features a large fast-focusing hand wheel at its base. (Photo courtesy Questar Corp.)

Be sure to check that an erector is available for the telescope of your choice; also find out its price, for some can be quite expensive.

Another problem sometimes encountered when a telescope designed for astronomy is used in bright daylight is that it may suffer from poor *contrast*. Contrast is a measure of how well two nearly equally bright or nearly equally colored surfaces are discriminated from one another. High contrast is best for nature studies—the image should appear crisp and have a vivid separation of hues. Low contrast gives an image that appears

The principle of contrast. *Top:* The moon during daytime. Its image has low contrast relative to that of the sky. *Bottom:* The moon at night. Its image has high contrast. The moon's brightness didn't change, but the sky's did. (Photos by Dennis di Cicco)

washed out, as though the scene were being viewed through a strong haze. Telescopes designed for astronomy sometimes do not have adequate internal baffles to block extraneous light rays; these baffles are needed in daylight to maintain good contrast. (Binoculars can also have inferior contrast, so check a prospective model against others to make sure your choice measures up.)

Sunglasses Before going on, I want to take a few paragraphs to describe sunglasses, our most familiar optics for outdoors, and to describe, among other things, how their tints enhance contrast. As any landscape photographer will testify, if you're working in black-and-white the standard trick is to put a yellow or red filter in front of the lens to darken the sky and make clouds jump out of the blue background. In other words, the filter increases contrast. The same principle works for your eyes. On cloudy or hazy days, a yellow or amber tint will markedly improve contrast. Traditional green seems to be the most useful under conditions of moderate brightness—maybe that's why this color is so popular. In bright sunlight gray is especially good at cutting glare while increasing contrast; this tint also has the advantage of best preserving the natural colors of the scene. (Some sunglasses, the so-called photochromics, darken according to prevailing light levels and therefore deliver a constant brightness to your eyes.)

A lot of confusing debate (and advertising) has recently revolved around the superiority of certain sunglass tints for blocking ultraviolet light, thereby presumably helping to prevent cataracts. The simple truth is that all glasses block ultraviolet light; it's the material that does the absorbing, not the color. If you wear spectacles in your daily life, be they glass or plastic, you are blocking out from 50 to 99 percent of the incoming ultraviolet rays. Contact lenses offer the same protection.

Our landscape photographer also knows that a polarizing filter can produce contrast effects on color film that equal those a red filter creates on black-and-white. And polarizers, just like tints, work with eyeballs. They perform best when dealing with reflected light; after all, it's the reflecting surface that does the polarizing. So, if you're going to view a lot of water, ice, or bright metal surfaces, a polarizing eyeglass is the best choice.

There are also glasses that are designed to reduce glare. This type of specially prepared glass is often marketed under an "anti-reflective" label, particularly as a protection from oncoming headlights for nighttime drivers. In fact, any lens —your ordinary eyeglasses included—can be coated with layers of metallic oxide to do this job.

Sunglasses that clip to the outside of prescription lenses have a drawback for users of binoculars and spotting scopes, for their thickness adds to the distance between your eye and the eye lens of your instrument. This means that you will need to have a few millimeters of extra eye relief in your system, otherwise you will start getting a constricted field of view. Fortunately there is an alternative—tinted glasses that fit *inside* your spectacles. Since they do nothing to change the distance between your eyeball and the eyepiece, your usual field of view will be preserved.

Magnification

Let's get back to spotting scopes and the question of how much magnification is reasonable. Applying less than 20 power would underutilize such an instrument. On the other hand, you can rarely employ more than 40 power because of blurring due to tripod vibrations or shimmering due to "heat haze." The latter is actually a misnomer, for this effect becomes troublesome whenever there is a large temperature difference between the air and the land or water over which you are viewing. Thus, the shimmering will be just as objectionable when the air temperature is ten degrees and the land thirty degrees as when the air is ninety degrees and the land seventy degrees. Heat haze is just another manifestation of the blurring you see when you look down a hot roadway or across a rooftop on a summer day.

Now the issue becomes: How big should the objective lens of a spotting scope be? This topic harks back to the discussion of exit pupils. In this context it is important to realize that the exit pupil of a spotting scope can be rather smaller than that of a binocular. This less-demanding criterion results, in part, from the fact that scopes are normally mounted on a tripod or equivalent, which provides stability and aids in centering your eye on the eyepiece. Since you will want to use your scope at magnifications between 20 and 40 (perhaps up to 60 on very rare occasions), a 60-mm aperture seems about ideal. It will deliver an exit pupil of 3 mm at 20 power, or 1½ mm at 40. Thus, in bright daylight, when your eye's pupil is about 2 mm in diameter, you will have full image brightness up to 30 power. The reason for not going much below 60 mm rests solely on the need for obtaining a sufficiently large exit pupil for daytime viewing. (Because of their inherently small exit pupils, spotting scopes become quite inefficient under dim lighting conditions.)

"But why shouldn't I buy a larger scope?" you ask. "One with 80 or 90 mm of aperture so that the daytime image will stay at full brightness even at 40 power?" The answer is

Old and new ways to change magnification. *Top:* The classic Bausch & Lomb Balscope Sr. with its now-unfashionable bulky rotating turret holding three fixed-power eyepieces. *Bottom:* The contemporary Bushnell Spacemaster equipped with a zoom eyepiece. (Photos by author)

threefold, each part based on the same thread of logic. First, an 80- or 90-mm spotting scope would be about twice as bulky as one with a 60-mm lens. Second, it would also likely cost much more. Third—a little-appreciated point—the larger aperture would tend to increase the conspicuousness of heat haze. Nevertheless, it's hard to argue against the legendary Questar (89 mm) or the recent success of Kowa's 77-mm scope.

Unlike binoculars, which normally have only one magnification, most spotting scopes allow you to change the power. There are three basic ways to do this: a zoom system (either built into the scope or available as a special eyepiece); a rotating turret having several eyepieces; or a set of interchangeable eyepieces. This is also the order of convenience in field use, but it is not necessarily the order of quality. To judge the latter, you will have to test each scope and each eyepiece or zoom system individually.

Manufacturers attempt to make their eyepieces (or zoom systems) *parfocal*. That piece of technical jargon merely means that when you change magnification the image remains in sharp focus. Unfortunately, no spotting scope accomplishes this task perfectly, though some succeed much better than others. Inevitably, after changing the power, the focusing device has to be tweaked a bit to get the crispest possible image.

Making such a fine adjustment is not a problem as long as the focusing control is conveniently placed and has a firm, quick action. I especially prefer a small knob mounted either on the barrel of the scope or at the eyepiece end. I don't like focusing devices that require me to turn a ring encircling the scope's barrel—these usually require a lot of force, which results in a bouncy image.

Some spotting scopes, such as this Swift 20- to 60-power Lynx, allow you to change magnification by means of internal zoom systems. This instrument employs the so-called Schmidt-Cassegrain optical design and is here set up for photography. (Photo courtesy Swift Instruments Inc.)

Additional Features

Other than such basic qualities as sharpness, compactness, sturdiness, and design, there are a few other spotting-scope features to look for. Let's examine six important, though secondary, ones that have proved their usefulness time and again.

Rain/Sun Guard I wish every spotting scope had a 2-in. or longer projection of tubing in front of the objective lens. (Especially convenient are instruments with built-in sleeves that can be extended when needed.) Scopes are often used in very foul weather, such as when looking for ocean birds during storms, and if the lens becomes covered with water, you can't see anything. A good rainguard can also protect and preserve the lens. If too much moisture accumulates on the lens, it might seep between the glass elements and perhaps fog the image for an entire day. Even after the moisture evaporates it can leave a permanent light-reducing stain on the lens. An extension is useful on fair days too, for it will block extraneous sunlight that can cause glare on your image. Photographers are keenly aware of this problem and routinely use lens shades to prevent unwanted glare.

Protective Caps A spotting scope should have hard caps to cover both its front lens and its eyepieces. Too many times have I seen good optics accidentally damaged when the excitement of the quest overshadowed the prudent care of equipment. Your optics are particularly vulnerable when you travel with companions. A lot of scopes and tripods bouncing freely in the trunk of a car is almost certain to lead to disaster.

Internal Focusing There are two ways to focus a spotting scope, externally or internally. External systems have parts

Unlike spotting scopes from the United States and Japan, those from Europe often feature draw tubes, which form a compact package when the instrument is closed. *Above:* Germany's Optolyth 30 × 75 set up on that firm's Uni-Pod, which features a clamp and swivel head. (Photo courtesy HP Marketing Corp.)

Identically sized as the scope on the previous page, this Austrian model is made by the firm of Swarovski. Both scopes are armored with rubber. (Photo courtesy Swarovski Optik.)

that you can see move when you change focus, just like the lengthening or shortening of a pair of Porro-prism binoculars. Internal designs accomplish the focusing task by moving optical elements you can't see. Most scopes manufactured today use the latter method. As in roof-prism binoculars this is preferable, for the optical system remains better sealed against dust, dirt, and especially moisture.

Straight-Through Viewing A number of manufacturers offer models in which the eyepiece is tilted at an angle relative to a line passing through the center of the objective lens. Sometimes this angle is ninety degrees—you look directly down toward the ground to see an object on the horizon—sometimes it's at forty-five degrees. Often the eyepiece is merely offset from the instrument's optical axis—you look parallel to that axis. This configuration is particularly common when a Porro-prism erector is used to give a right-side-up image. In my opinion, for terrestrial use it is best to have the eyepiece placed as closely as possible in line with the center of the objective; in other words, along the optical axis. This arrangement makes it much easier to get your target in the field of view. An exception to this rule is astronomical observing, where the scope is often pointing upward at a very steep angle. In this situation a tilted eyepiece offers a real advantage by allowing for a more comfortable posture while helping you avoid a real pain in the neck.

An example of a versatile camera adapter is this one for the Bushnell Spacemaster. Note how the extendable portion of the barrel is calibrated: It gives effective focal lengths and f-stops for 15-, 20-, and 25-power eyepieces. (Photo by author)

Mounting Fixture Any spotting scope should have a *very* substantial bracket for attaching the instrument to a tripod or other supporting device, such as a car-window mount or gunstock. By the way, when attaching your scope to a mounting, take care not to cross-thread the screw. Repairs can be extremely expensive!

Photographic Adapters Naturalists often want to take photographs of their subjects. So you should know that many popular spotting scopes have optional and inexpensive adapters that turn these instruments into telephoto lenses. But before you get too excited, remember that most spotting scopes and all binoculars are not substitutes for true telephoto lenses. The latter have much more sophisticated optical designs, particularly to give a sharp image across the entire width of a film frame.

Yet, by coupling a scope to a single-lens-reflex camera you can create a powerful tool for recording some special moment with nature or to document the sighting of some unusual critter. Such photography has never been easier thanks to amazing recent improvements in fast fine-grain color films and cameras that think for themselves. If this kind of activity appeals to you, make sure that the manufacturer of your scope of choice can provide a photography adapter. (Incidentally, any large camera store should have publications dealing with telephotography.)

Five popular spotting scopes with apertures from 50 to 89 mm are listed in Table B; its arrangement is similar to the table for binoculars in Chapter 5. Again, I've limited the selection to a few representative models. Where appropriate, below the manufacturer's name is that of the particular scope.

Celestron's Ultima 8 is a state-of-the-art mirror-lens telescope for astronomical viewing and photography. A built-in drive system follows the stars in their daily motion across the sky. (Photo courtesy Celestron International)

The objective-lens diameter is listed next, followed by the magnification (or range thereof) and the corresponding exit-pupil diameter(s). The fifth column gives the diameter of the field of view at a specific magnification. Indicated in the sixth column is whether the power is fixed (F) or can be changed by switching eyepieces (S) or by zooming (Z). The next two columns give the telescope's weight and suggested list price for 1989. Accompanying notes highlight some available options and special qualities.

The second part of this table features two astronomical telescopes. They are kept separate because they are highly specialized pieces of equipment and would rarely be used for terrestrial viewing (though superb nature photographs have often been taken with such telescopes, which act as extremely long-focus camera lenses). I've selected two state-of-the-art units, one a mirror-lens system and one a "traditional" refractor. Much of the technical data is the same as for spotting scopes, but note that the magnification column has been replaced by "Maximum Magnification," which is the highest power you can reasonably expect to use on the telescope. (I've adopted the long-standing convention of 50 power per inch of aperture as the maximum.) Very high powers are often employed in astronomical observing, and, since the clarity of the image is often determined by the steadiness of the Earth's atmosphere rather than by the optical quality of the instrument, you might not always be able to exploit the maximum listed. Also, the exit-pupil column has been replaced by f, the *focal ratio* of the telescope. This is the same f-number that is familiar to photographers and represents the speed of the system, that is, its focal length divided by the aperture. Finally, the weight column is now given in pounds, not ounces as previously. Astronomical telescopes and the mountings necessary to support them are truly heavyweights!

Table B: Five Exceptional Spotting Scopes for Nature Study

Mfg. and Model	D (mm)	Mag.	EP (mm)	FOV (dg)	M	WT (oz)	$	Notes
Leupold	50	20	2.5	2.2	F	18	459	1
B&L *Spacemaster*	60	22	2.7	3.1	S	36	402	2
Swift *Lynx*	65	20–60	3.2–1.1	2.2 @ 20x	Z	42	715	3
Kowa *TSN-4*	77	20	3.9	3.4	S	52	1,340	4
Questar *Field*	89	50–80	1.8–1.1	1.5 @ 40x	S	48	1,890	5

Two Exceptional Telescopes for Astronomy

Mfg. and Model	D (mm)	Max. Mag.	f	FOV (dg)	M	WT (lb)	$	Notes
Astro-Physics *StarFire*	127	250	8	0.8 dg @ 56x	S	52	3,289	6
Celestron *Ultima 8*	203	400	10	0.7 @ 68	S	66	3,325	7

D	Diameter of objective lens
Mag.	Magnification or range available
EP	Exit-pupil diameter for the magnification given (or range thereof)
FOV	Field of view diameter
M	Indicates whether the magnification is "fixed" (F—that is, only a single power is available) or whether it can be changed by switching (S) eyepieces or by zooming (Z).
WT	Weight

$	Manufacturers' suggested list price, 1989
Max. Mag.	Maximum magnification that is generally useful; 50 per inch of aperture adopted here.
f	Focal ratio of system; the focal length of the objective divided by the diameter of the objective.

Key to Notes:

1. Armored version available.
2. With wide-angle eyepiece; other eyepieces with magnifications from 15 to 60, also 15–45x zoom. Very rugged and compact. Most popular scope among bird-watchers. Photo accessories.
3. Sophisticated mirror-lens system. Photo accessories.
4. Fluorite lens elements. Wide-angle model; other eyepieces with magnifications from 20 to 60; also 20–50x zoom. Bayonet eyepiece mounting for quick change. Photo accessories.
5. Mirror-lens system; base price. Extreme versatility, visually and photographically. Reverses images left for right. Bulky in field use. Useful for astronomy.
6. Many options available, such as electronic drive to automatically track stars; also photographic accessories. The model cited does not have a drive.
7. Mirror-lens system. Wide variety of magnifications available. Photo accessories. Electronic drive system follows stars.

9

Spotting-Scope Supports

Because of their high powers, spotting scopes need some kind of support other than hands alone. There are three popular ways to do this: using a tripod, a car-window mount, or a shoulder brace. Since the tripod is an essential piece of hardware during almost any outing, we'll discuss it first. My remarks apply to full-size tripods that are carried in the field, not the so-called table-top models that stand only a foot or so high and are widely used by target shooters or for looking out picture windows. At the end of this chapter we'll look at some alternative ways to support a scope.

The problem with tripods, of course, is deciding which one to buy—there are dozens of manufacturers from which to choose (see Appendix B for some addresses). It seems to me, from the sad assortment of tripods I've seen in the field, that their selection is often an afterthought. I can imagine someone getting his or her first spotting scope as a present and then suddenly realizing that it has to be mounted on something. A lightning trip to a local camera or department store ensues, followed by an equally quick scan for a tripod that looks familiar—something like Outdoor Joe's. The purchase is soon made, and that's that. After all, who can get excited over a tripod; it's just a widget to make the *real* equipment work. Thus another shaky, cumbersome, or otherwise ill-suited support for a first-rate scope joins its cousins on some windswept shore.

Not surprisingly, the quality, utility, and cost of tripods differ greatly. The ideal unit would be lightweight, compact, and quickly set up. But herein lies the rub. The lighter the tripod, the more likely its stability has been compromised, and a shaky piece of equipment is of no use whatsoever. Compactness usually means lots of leg joints, and in time these may become either too loose or too arthritic.

It is important to remember that most tripods have been designed for cameras, not spotting scopes. Therefore, you have to look around carefully to find a model that is durable, stable, and, most of all, convenient to use. As in choosing binoculars or spotting scopes, try out your friends' equipment first—the tests are free and the variety superb. In particular, make sure your sampling includes tripods that have seen a lot of dirt, water, salt, and sand. It's amazing, but the performance of equipment in the field after months or years of use often falls far short of what it was when the unit was brand new!

When selecting a tripod, you should keep four features in mind—these are the ones that you will find really useful.

Stability This is paramount and depends on the tripod's design, weight, and the materials used in its construction. For outdoor use, you should avoid so-called open-rail models, which are usually made from U-shaped sheet aluminum. These tripods do not effectively absorb ground or wind vibration or knocks imparted by the user. At the opposite extreme are traditional wooden tripods, which can be very sturdy but generally are too heavy to lug about on a day's field trip. For most people the best choice will be a tripod formed from tubular aluminum with relatively thick walls.

When you talk with a salesperson at a major camera store, be sure to explain exactly what you intend to do with the tripod and indicate the kinds of situations and conditions you anticipate encountering in the field. If you already have a spotting scope, it's a good idea to take it with you to the store so you can mount it up on any tripod candidates that catch your eye.

A good test of stability is to extend the tripod legs fully (but not the center post if the tripod has one) and then push down quite hard on the head (the place where you attach the spotting scope). If the legs bow under this pressure, the tripod is too weak. A second test is to give the tripod a moderate whack with your hand while looking through the spotting scope. A good model will shudder for a second or two and then become still. A flimsy or ill-designed tripod, on the other hand, will tend to sway back and forth for a much longer period of time.

Flip-Lock or Screw-Down? Usually there are two choices when selecting a tripod design: the flip-lock, where the flick of a lever releases or engages the legs; and the screw-down, where the legs are secured by means of a friction ring that may be turned. I don't know of any study that has indicated one design

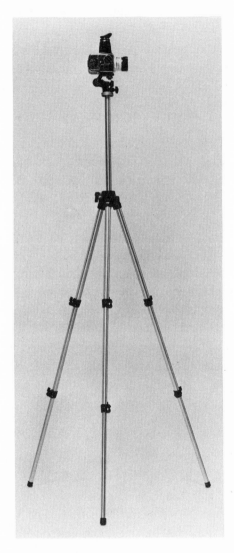

Bogen, a leading manufacturer of quality tripods, offers this general-purpose "3020" flip-lock model. Each leg has three different click-stop positions to set the spread angle. Note the interesting tripod head—no handles, just knobs. This unit would not be a good choice for most naturalists, but it illustrates one extreme option. Bogen also offers many conventional heads. (Photo courtesy Bogen Corp.)

to be fundamentally better than the other. However, flip-locks are clearly superior for rapid set-up. Their locking levers are quicker to operate than screw-type rings, and the leg tubes can be nested inside one another with wider tolerances than the legs of screw-down designs.

"So what?" you say. "Who cares about tolerances?" Well, here's another subtle connection between manufacturing practice and field performance. In particular, the legs of some screw-down tripods are so tightly fitted that they must be compressed forcefully by hand to overcome internal air resistance. (A small, carefully drilled hole at the base of each leg might cure that problem, however.) On the other hand, gravity alone will suffice to open and close most flip-lock models. Furthermore, in cold weather, wearers of mittens or gloves will find it difficult to tighten a screw-down tripod leg; flip-lock levers are much easier to operate.

Tripod Heads Many manufacturers of tripods offer a variety of heads, but most of these are intended for special circumstances encountered in photography. Spotting scopes have much simpler requirements. For example, only vertical and horizontal motions are needed for pointing. The tripod head should be rather heavy, to suppress vibrations as you move the scope around. Also, it should sit close to where the legs come together—putting the head atop a long extension, such as offered by a center post, only decreases stability. Finally, when examining a head, make sure its various handles and knobs are of adequate size and conveniently placed (many heads have a handle that will try to skewer your neck!).

Almost unknown among outdoor folk is the so-called ball-head design, which offers an interesting alternative to the omnipresent three-axis heads, which tilt vertically and side-

This Star-D tripod has been set up for maximum stability in both stand-up and sit-down situations. This particular design incorporates friction rings to secure the legs. (Photos by author)

ways and rotate horizontally. These devices lack the familiar porcupine array of handles and rely instead on compact knobs or levers to adjust tension on a spherical bearing that holds your scope. Ball heads are very strong and a favorite among professional photographers who need to follow moving targets. Unfortunately, high-quality models cost as much as most spotting scopes.

Tripod Size You should make certain that the tripod will stand tall enough so that the spotting scope will be at eye level when *only* the tripod's legs are extended. Less obvious but equally important is that the tripod should also collapse to a height that will allow you to scan comfortably while sitting on the ground or, for a handicapped person, in a chair. I find sitting on the ground particularly enjoyable—yes, even in winter with a "seater-heater" under my rump—for the short tripod is as rock-stable as my position is restful.

The above discussion about tripods applies only to typical spotting scopes, those with objective lenses having diameters of 50 to 70 mm or so. "Superscopes," such as Questar's 3½-in. (89 mm) are heavier and require more solid support. One leader among manufacturers of massive tripods is Linhof. The worst nor'easter will convince *you* to quit before one of these units gets the message. The price to be paid—in addition to several hundred dollars—is weight. Such heavy tripods will severely limit your maneuverability, for their main function is to supply a stable platform for photography or visual observation under *very* high power, such as is often used in stargazing.

Tripod Alternatives Among the most popular alternatives to a tripod is the car-window mount. These are especially nice if you can get to a convenient vantage point, such as a dike around a waterfowl compound or a bluff overlooking an ocean

A handy spotting-scope accessory is a car-window mount. When selecting one, make sure the handle doesn't project so far back as to interfere with viewing. (Photo courtesy Swift Instruments Inc.)

dotted with multicolored sailboats. They also offer a great advantage to handicapped viewers, who can take in the sights from the comfort of the car, or to anyone who wishes to observe in extremely cold or foul weather. A window mount is much more than a lazy person's convenience—a car acts as a blind and permits you to get much closer to birds and animals than you might otherwise be able to do. Furthermore, when exploring wide, unobstructed areas, such as oceans or prairies, a window mount can be extremely efficient for scanning a lot of territory. Several spotting-scope manufacturers offer these handy devices, and there are also a few generic mounts that can be purchased through independent suppliers.

The gunstock mount or shoulder brace is another alternative support, though one that requires rock-steady arms. To use one effectively you need to form a triangular configuration— arm, gunstock, shoulder—and be able to tighten up on this assembly without causing trembling or a wandering aim. Gunstocks, which are available from large camera stores and specialty houses that advertise in photography magazines, are particularly useful on ocean-going ships or smaller coastal vessels in relatively calm seas. With practice, a bird- or whale-watcher with a modest-power scope can nicely compensate for the rocking and pitching of a boat.

Two considerations are especially important for those who wish to get really close-up looks at animals at sea, especially from a small boat. First, use a scope having about 20 power. Much less magnification won't give views all that much better than your 8- to 10-power binoculars. Greater magnification will simply be too much to hold steady. If a wide-angle eyepiece is available for your scope, use it; the animals will be found more quickly and can be kept in the field of view more easily. Second, choose a short-barreled scope so that its "lever

Although suffering from many trips around the world, this lightweight, compact metal-and-plastic gunstock mount performs like new after fifteen years of use. Such a support is especially useful when joined to a short-barrel spotting scope or telephoto lens of 500-mm focal length or less. This Rowi model has three adjustments: horizontally forward and back along the top rail for the scope or camera; horizontally forward and back for the extension carrying the butt, to accommodate arm length; and vertically up and down so the butt can match shoulder-eyepiece separation. The unit pictured had been retrofitted with a heavy-duty camera cable release held in place by electrical tape; it has since been retrofitted again to accommodate an electronic shutter release. (Photo by author)

Bogen's model "3018"
monopod features three sections
and extends from 27 to 65 in.
It weighs only 29 oz. (Photo
courtesy Bogen Corp.)

action" will be minimized as the boat bobs about.

The final and least-used alternative support is the *monopod*, which looks exactly like one leg of a tripod. Since it has only one leg, it must be supported by your hand, so some swaying and shaking results. Yet these devices can be set up very quickly, and when used with relatively low-power instruments they perform surprisingly well. And monopods are very compact and lightweight, attributes long-distance travelers and backpackers should particularly welcome.

Cleanliness
and Care

Optical instruments have to be kept clean if they are to perform at their best. Yet I'm always astonished to learn how many of my friends who own first-class binoculars and spotting scopes ignore this obvious fact. It's not as if they weren't told the right things to do; there is perfectly good advice in every instruction booklet that comes with new equipment.

The two principal enemies of optics are ordinary dirt and scum (such as fingerprints or other grease, smoke, and residues from evaporated moisture). I've been told that a really filthy pair of binoculars can snuff out up to 50 percent of the light that strikes its front lens. If you're really unlucky you might even start a fungus farm. A more insidious problem stems from the fact that scum and dirt scatter light very efficiently, and the result will be an image with dramatically reduced contrast. Instead of crisp, sharp details with vivid colors, all you see is a dull, blurry monochrome. Like growing near-sightedness, such a degradation usually increases so gradually that it goes unnoticed. Often, a person's first clue that something is wrong comes when they try someone else's new equipment. Then you hear this complaint: "You know, my binoculars used to give super images, but now everything is hazy and dull." They speak as though the optics have worn out; they don't realize that a five-minute rehab could probably make them as good as new.

You should routinely check your optics after each outing —particularly the eyepiece lenses, which tend to gather everything from skin oil and eyelashes to cookie crumbs and cigarette ashes. Do this especially when the optics have been subject to harsh conditions such as dust or salt water. Cleaning is a simple procedure, and I'm going to go through the steps because I know a lot of folks don't take the time to read their instruction booklets. (Some experts might say my technique is too cavalier; maybe so, but I've yet to damage any optics.)

Begin by blowing or carefully brushing off any obvious specks of dirt, and then clean the objective lens and eyepiece with a breath or two and a *very* gentle rub with toilet tissue. (Don't use a cloth made of synthetic fibers or one treated with silicone, and stay away from Kleenex-type products that leave a lot of lint.) If the scum is particularly bad, rub the lens surfaces lightly with a household window-cleaning fluid (I like Windex) *after* it has been applied to the tissue. If you pour the liquid directly onto the surface it may "bleed" between the lens elements and cause fogging or a stain. Don't use rubbing alcohol, for it will leave a residue on the surface of the lens. Camera stores can supply special lens-cleaning fluids and lint-free wiping papers, but these don't seem to work much better than products found around the home. To test whether lenses are really clean, breathe hard on them. If the moisture evaporates evenly without showing any streaks or swirls, you've done a good job.

One well-known bird-watcher whom I respect advises purchasers of new binoculars to throw away the instrument's protective lens caps and case. He argues that binoculars are meant to be used and should be ready at any instant. Although some manufacturers who read such advice will cringe, I can agree with the motive—at least insofar as the lens caps are concerned. They do take too long to remove, and if the adjustable eyepiece isn't secured by click-stops or some other means, it's almost sure to come away misfocused. Furthermore, the lens caps always seem to fall off when they're not supposed to. Of course, that's an easy way to get rid of them.

The binocular case is a different matter. At the start of the day, a pop of the top or a zip of the bag will get your binoculars into your hands soon enough. And when the day is done, there is no good reason for not putting the binoculars back into their case before stuffing them under your car seat or hanging them by a window. A lot of dust and scum will thereby be kept off the optical surfaces at no inconvenience to you.

Cases come in two basic kinds—hard and soft—and serve different needs. Hard cases offer real protection against physical damage from shocks, such as when a baggage handler tosses your binoculars across a receiving room. (Seasoned travelers avoid this problem by always packing their binoculars in carry-on luggage.) Soft cases, which are particularly popular for roof-prism models, don't offer the same protection, though they are more convenient than hard cases when you stuff your

West meets East with these 10 × 40s. *Left:* A pair of Zeiss
(Oberkochen, West Germany) binoculars with a soft case. *Right:* An
aus Jena (East Germany) model with a hard case. (Photo by author)

binoculars into luggage. Nevertheless, a soft case is still adequate to ward off dirt and scratches from sharp-edged objects.

When traveling, especially internationally, be aware that premium leather advertises quality goods to rip-off artists. It's also a wise policy to cover the manufacturer's logo with tape—those of Bausch & Lomb, Leitz, Nikon, and Zeiss are particularly eye-catching to people who would like to lighten your load!

Try to care for your binoculars or spotting scope as you would a fine camera. Most instruments can take a good deal of knocking about and misuse; just be careful not to abuse them unnecessarily. Fortunately, the major manufacturers provide excellent repair services, though some may require weeks or months to do even the simplest job.

I'm reminded of an anecdote to illustrate once again, if another example is needed, that a little knowledge can be a dangerous thing. A fellow accidentally dropped his brand-new binoculars into salt water. Knowing that the salt would cause corrosion, he first let them drain and then plopped the binoculars into a can of used motor oil! After receiving this unprecedented package a few days later, the manufacturer was nevertheless able to make the glasses as good as new. I'm sure that by now someone has managed an even better way to test a warranty.

But there will be times when bird-watchers, sportsmen, and sailors *have* to abuse their equipment. Once in Africa it became so hot that the grease in my binoculars flowed like water over the prisms. (You can get the same result by leaving your binoculars in an overheated car!) On a much cooler occasion in the Arctic the grease became so stiff that I barely could change focus. Few manufacturers design their instruments to cope with all bizarre situations.

Tips for Testing Equipment

Even new equipment from a first-class supplier can be out of whack when you first hold it in your hands. Once a binocular or spotting scope leaves the manufacturer it might be tossed around by shippers, truckers, and stock clerks. And even after an instrument has made it to a showroom in good condition it can be dropped or otherwise abused by a customer who may then return it surreptitiously to a display case. So when you visit a store or receive new optical equipment in the mail, take the time to make a few quick checks to satisfy yourself that the instrument's performance is up to snuff.

First, look for obvious signs of mishandling, such as a tattered box or a crushed Styrofoam liner. Then check for

111

nicks or dents on the body of the instrument, and don't forget to look at the objective and eyepiece lenses for scratches, dirt, or fingerprints. Now, with your eye six to twelve inches away, look at the sky through the *front* of the objective lens—just as you did as a kid to make the world become small. Then slowly swivel the instrument so that you get a good view of its insides. Look for any cracks, chips, or dirt on the lenses and prisms.

How about the focusing mechanism—does it move smoothly or does it bind, slip, or just act less than right? For binoculars, the rotation of its halves around the center post should be smooth and easy, as should the turning action of the adjustable eyepiece. Especially important when checking Porro-prism binoculars is to make sure the *bridge* connecting its two eye-pieces doesn't rock appreciably. If it does, uneven pressure from your eye sockets may cause the image to go out of focus. Test for such sloppiness by holding the binoculars as you normally would in the field but with your hands moved up toward the eyepieces. Now apply moderate upward and down-ward pressure with your thumb and forefinger as you alter-nately try to push one eyepiece, and then the other, in and out. If the eyepieces rock by more than a millimeter (⅟₂₅-in.), you might consider getting another pair.

The ability to produce a sharp, crisp image is paramount for any instrument. If the individual optical components are of poor quality, nothing can be done. But even good optics can deliver inferior performance if the prisms and lenses are improperly aligned (sometimes described as "out of colli-mation"). Misaligned binoculars cause eyestrain and head-aches because the user's brain tries to fuse the two separate images. This situation becomes especially bad when our eyes are forced to rapidly adjust, readjust, and readjust again. Bird-

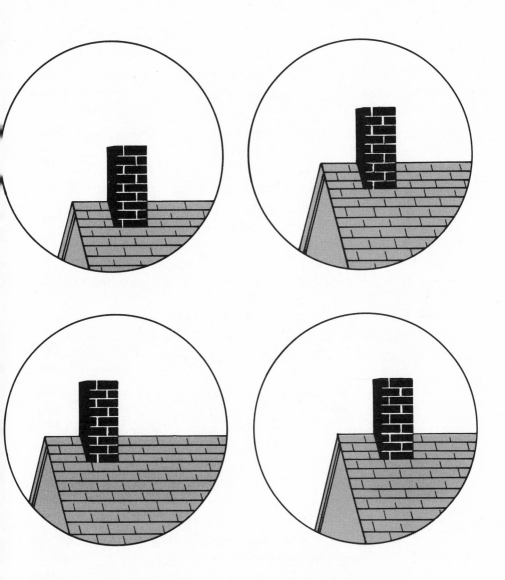

When the two halves of a binocular are out of adjustment and do not form identical images, the instrument may be unusable (at worst) or will cause eyestrain (at least). The top pair of sketches demonstrates vertical misalignment, where one image is higher in the field than the other. The bottom pair shows a horizontal, or side-to-side, mismatch.

watchers, especially, should take note, for this is exactly what their eyes must do hundreds of times during a day's field trip.

Here's a quick test to see whether the two halves of a binocular are in *vertical* misalignment—when one half shows an image above or below the other half. (This, in fact, is the most common type of misalignment.) I like to use the edges of the clapboards on my house. First, *slowly* move the binoculars away from your eyes, being sure to keep an imaginary line through the centers of their objective lenses parallel to the clapboards. For the first inch or so the images from the two halves will remain fused into one. But as you continue to move the binocular farther away, the image will split into two, one in each eyepiece, and their fields will eventually constrict to show only a single clapboard. If the binoculars are aligned properly, the inner edge of the clapboard in the left image will match up exactly with the inner edge of the same clapboard in the right image.

Despite demanding unnatural muscular movement, human eyes can fuse images that are slightly misaligned vertically. Individuals, however, differ greatly in their abilities to accomplish this task. But whether you can succeed or not, *any* vertical misalignment will cause eyestrain, particularly if the observation is protracted or repeated frequently.

Perfect *horizontal* alignment is also best, of course, but it is not always absolutely necessary. The reason is the marvelous ability of our eyes to "fix" small mismatches in our vision. Let's say the images from the two halves of the binocular are slightly horizontally misaligned *inward*. Your eyes will readily compensate by "toeing-in" to cause the images to fuse. Eyes make this adjustment easily and naturally. They do it all the time, otherwise you couldn't make the transition from watching a baseball go out of the park to finding the place on your scorecard to mark the run down. The opposite case—when the images are horizontally displaced *outward*—is quite a different matter. Humans don't have the facility of lizards to "toe-out." We simply cannot fuse such discordant images.

There is a simple test for judging horizontal mismatch. First, place the binoculars on a stable surface and focus them on some distant object that covers only a portion of the field of view. A chimney on your neighbor's house, for example, is excellent. Then, cover the objective lens of the right half of the binocular and mentally note the position of the chimney relative to the edge of the field. Now reverse the process and cover the left half. Is the chimney in exactly the same place in the field? If not, you've got a horizontal mismatch.

The one thing everyone anticipates when buying a new spotting scope or pair of binoculars is a sharp image. There are several things you can (and should) do to check out its quality. The first is to look at something you can later examine close up. How about a brick wall? Do you clearly see each bit of surface texture—and do you see it equally well across the field of view, from the center to the edge? If the instrument has poor optics, individual grains in the mortar or pits in the bricks will appear soft-edged all across the field. More commonly, the image will be sharp in the center and then will soften toward the edge. Virtually all binoculars and spotting scopes show this latter effect to some degree. Ask yourself this question: Is the image sharp nine-tenths of the way to the edge, three-quarters of the way, one-half? The quality of the instrument can be judged directly by the size of the fraction you choose.

There are also more direct and quantitative ways to test optical quality. Place this book twenty-five feet away (a foot more or less won't make much difference) so that the chart on the back cover is brightly illuminated by sunlight. Now, if you have normal 20/20 vision, you should just be able to separate the black-and-white bars at the top of the chart with your naked eye. The bars might not appear really distinct; rather, you might only "sense" their presence because your eyes will be working at their absolute *resolution* limit—the ability to separate two adjacent objects. (In technical terms, this "ultimate" resolution of the eye is about 2 minutes of arc, a 30th of a degree or a 10,800th of a circle. Dim light or low contrast will reduce your resolution limit to 5 minutes of arc or less.)

This chart, of course, can also be used to test binoculars and spotting scopes. Since the bars at the bottom are ten times closer together than those at the top, the bottom bars should just become visible when viewed from twenty-five feet through 10-power binoculars having excellent optics. If you want to test a spotting scope by using the bottom bars, move the chart fifty feet away for 20 power and one hundred feet for 40 power. For these tests you should firmly support the binoculars or spotting scope on a tripod or by some other means, such as on a bench, box, or windowsill.

The chart can also be used to compare the relative performance of different instruments. In this case, all you need do is look at the chart through one spotting scope or binocular— say, the one you've been thinking of replacing—and compare its performance to that of another. Of course, both instruments must have the same magnification, and the test must be carried out with the chart at the same distance. (If the magnifications differ, the chart will have to be moved to compensate.) Such a head-to-head test will quickly separate the sheep from the goats.

All of these tests should begin by placing the chart in the center of the field of view. That's where any optical instrument should give its best performance, and that's also where you will usually place your subject. But remember also to check how well the resolution holds up when you swivel the instrument to move the chart toward the edge of the field, for that's where quality optics really show off compared to lesser ones.

A star, which acts like a point source of light, provides the acid test for evaluating binoculars and spotting scopes. With the instrument held steady by a support, focus on a bright star placed right at the center of the field of view. The very best optics will produce a tiny dot with little or no fuzz around its edges. Certainly no bright spikes or flares should be visible, nor, in roof-prism models, should the image appear split or butterfly-shaped.

If you move the star toward the edge of the field of view, its image will soften just as did the bricks and mortar mentioned earlier. It will also become elongated. Virtually every optical instrument used in nature study will show this effect. The question is: How elongated? At the very edge of the field of view, something like twice as long horizontally as vertically would indicate a really good piece of equipment. But if the image becomes a streak rather than a dash, or dissolves into a blob or some other weird shape, you might reconsider your selection. As we discussed earlier, some people use the edge of the field of view and some don't (eyeglass wearers, particularly). You alone have to judge how much distortion is tolerable.

Spurious colors seen in a binocular or spotting scope come in two varieties. The first, an overall tint across the entire field of view, is a product of the instrument's total design—from its glass to its coatings. To check for this kind of spurious color, just look through the instrument at a white surface, such as the side of a house. Then quickly look at the same surface with your naked eye. Do the colors closely match? The test is as simple as that. A scene viewed through the very best equipment, in my opinion, will look just as it does to the naked eye.

This is not to say that an unobtrusive tint is bad. As with sunglasses, tints that produce a "cooler" image actually seem to yield a desirable increase in contrast. But some binoculars and spotting scopes tend to make the image much yellower or "warmer" than in life. One of the worst examples I've seen involves the venerable 1960s-vintage Bausch & Lomb Balscope Zoom 60. Yet, except for its yellow cast, that instrument yields a superb image and remains a classic of optical and mechanical design. It was truly ahead of its time.

The second kind of spurious color should occur only near the edge of the field and therefore be inconsequential to most users. The test here is to look at something bright seen against a dark background, such as a light bulb in a room or the moon in a nighttime sky. Alternatively, you can look at something dark, such as the top of a pine tree silhouetted against a bright sky. As long as there is strong contrast between the subject and the background this test will work.

First, place the subject in the center of the field and focus carefully. You will probably see no color. Then, slowly move the subject toward the edge of the field. Do you see an extensive, *brightly* colored fringe develop? Purple or blue is especially common. If you do, the optics are affected by *chromatic aberration* and are not of the best quality. You can also use the star test described previously to evaluate this type of spurious color.

Little Things

Rain may dampen our enthusiasm for nature study or other outdoor activities, but it won't keep hard-core aficionados out of the field. Under such conditions a rainguard for binocular eyepieces is an invaluable aid. These gadgets usually slide up and down along the instrument's straps through slotted tabs. When purchasing a rainguard, make sure it will fit over the eyepieces when the binocular is set to your interpupillary distance—many don't. Also, if you wear eyeglasses, check to see that the rainguard will fit over the eyepieces when their rubber eyecups are folded down as they would be under normal use—again, many don't.

Cold should be no deterrent either, unless the binocular

These binoculars were photographed after they had been set to match the author's very ordinary interpupillary spacing. Note that the rainguard for the Zeiss instrument (*left*) covers the eyepieces nicely. The one for the Mirador, on the other hand, doesn't and would be nearly useless in the field. Don't assume the manufacturer's rainguard will work for you. (Photo by author)

straps stiffen to the point where they stab your nearly frozen face. That's the major problem with plastic straps, which also tend to wear out quite quickly, though only about as fast as leather ones. Wear in plastic straps is particularly insidious, however, often being confined to a crack that remains essentially invisible until the binoculars take off—like Newton's apple—straight for the ground.

An excellent alternative is a strap braided from synthetic fibers. Such straps are strong and pliable at all temperatures; all you need do is be sure the strap is appropriate for the weight of your binocular. Ultralight models can get by with thin, stringlike straps. Binoculars in the 20-oz range do nicely with narrow flat straps, and many manufacturers now supply this type with their instruments. The heaviest models should have wide flat straps. Remember, a properly selected strap will spread the weight of a binocular comfortably around your neck and will inhibit chafing, which can be a major problem with heavy binoculars.

Any strap can fail, so check it from time to time. Also, I strongly recommend that you check occasionally for wear in the strap clips and their attachment fixtures on the binocular body. Not only should the fixtures be substantial, they should be positioned to allow the binoculars to rest flat against your chest. If the fixtures are properly placed and the straps are adjusted to their proper length, the binocular should not bounce up and down when you walk, even briskly.

The so-called Kuban hitch, which is popular among photographers, prevents bouncing by holding the instruments around your waist as well as around your neck. This kind of hitch also provides a backup in case one strap breaks, and it prevents your binocular from smashing into something when you bend over. Unfortunately, Kuban-type hitches are made for heavier-duty use than is usually needed for binoculars. A nice lightweight alternative is the so-called Ministrap, which uses narrow woven cords that slip behind your neck and around your arms. I have become very partial to this innovation for everything except my heaviest equipment. Ministraps are available from Avocet Enterprises, 1875 North First Avenue, Upland, California 91786.

Finally, I am continually amazed to see naturalists with many seasons' experience trekking over dune and mud flat, desert and meadow, while hand-carrying their spotting scopes and tripods. When they do so, their balance is upset and they

The value of a Ministrap, here holding a Bausch & Lomb 8 × 42 Elite, is threefold. It distributes a binocular's weight more evenly than a conventional strap does, provides backup support in case of strap breakage, and prevents the binocular from swinging far forward when the user bends over. A so-called Kuban hitch performs similarly but is more bulky. (Photo by author)

may tire quickly; they also have only one hand free to grab their binoculars when some critter flushes. I always carry my tripod on a homemade shoulder sling. My first ones were fashioned from electrical "zip cord" slipped through a leather "pillow" to cushion the weight. These slings lasted for years and worked fantastically. My latest sling uses the same pillow, but I replaced the zip cord with surplus straps from binocular cases. Of course, camera stores have much neater versions of this underexploited accessory. But why spend money when you can custom-make one yourself in just a few minutes?

The Final Word

Until World War II, binoculars made in Germany—particularly by Carl Zeiss before the firm was split up into Zeiss (West Germany) and aus Jena (East Germany)—had the deserved reputation of being the best available. This status should not be surprising since the prism configurations used in today's binoculars were developed in that country around the turn of this century. Carl Zeiss also introduced such innovations as wide-angle eyepieces and antireflection coatings.

But the war took a heavy toll on the German optical industry, and while it was recovering Japanese optics began to appear in ever-increasing numbers throughout American markets. For many years these products were, in general, poorly made, and

Japanese optics acquired a sordid reputation. Nowadays, however, superb optical instruments are produced in Japan (as they are in Germany and several other countries). Government quality-control measures and demands from importers brought this change about. "Made in Japan" no longer carries the stigma it did in the 1950s and 1960s.

Nevertheless, inferior optics can be made by any nation. So I recommend that you do business with a reputable firm —not necessarily one with a big advertising budget—and remember that you usually get what you pay for. You should consider spending a little extra to buy top-of-the-line optics. After all, the added cost of even the most expensive equipment, when spread out over ten, fifteen, or twenty years, doesn't amount to much. If you use the equipment frequently, the gains in performance and your greater satisfaction with a high-quality product will be worth whatever premium you paid. It's like operating a car; you get the lowest total cost per mile by driving it as much as you can each year.

Remember: *There are good optics; there are cheap optics; but there are no good cheap optics.* However, virtually all optical products can be purchased from discount stores, specialty mail-order companies, or manufacturers for 30 to 50 percent off their list prices. Advertisements in photography and nature magazines (see Appendix C) are good sources of leads. Look around carefully, and then be sure the manufacturer's warranty is valid in the United States. In other words, avoid "gray-market" goods, products that do not have warranties valid in the United States.

No binocular or spotting scope will perform perfectly for anyone all the time. The trick in selecting your next instrument is to get one that works for you most of the time. Look at your habits. Do you explore thickets, open country, or the seashore? Are your subjects nearby or far away? Can you carry heavy equipment all day, or would something lightweight suit better? Be sure to talk with knowledgeable people as you gain experience and gather information. Only you can pick equipment that is best suited to your needs.

Appendix A:

Binocular and Scope Manufacturers

A brand name different from that of the distributor or man-ufacturer is in parentheses.

Astro-Physics (StarFire)
7470 Forest Hills Road
Loves Park, Illinois
61111
(815) 282-1513

Celestron International
2835 Columbia Street
Torrance, California
90503
(800) 421-1526

Bushnell/Bausch & Lomb
300 North Lone Hill Avenue
San Dimas, California
91773
(714) 592-8000

Europtik, Ltd. (aus Jena)
P.O. Box 319
Dunmore, Pennsylvania
18509
(717) 347-6049

Fujinon Inc.
672 White Plains Road
Scarsdale, New York
10583
(914) 633-5600

Kowa Optimed, Inc.
20001 South Vermont Avenue
Torrance, California
90502
(213) 327-1913

E. Leitz Inc.
Link Drive
Rockleigh, New Jersey
07647
(201) 767-7500

Leupold & Stevens, Inc.
P.O. Box 688
Beaverton, Oregon
97075
(503) 646-9171

Minolta Corporation
101 Williams Drive
Ramsey, New Jersey
07446
(201) 825-4000

Mirador Optical Corporation
4051 Glencoe Avenue
Marina Del Rey, California
90292
(213) 821-5587

Nikon, Inc.
623 Stewart Avenue
Garden City, New York

11530
(516) 222-0200

Questar Corporation
P.O. Box 59
New Hope, Pennsylvania
18938
(215) 862-5277

Pioneer Marketing & Research,
   Inc. (Steiner)
216 Haddon Avenue, Suite 522
Westmont, New Jersey
08108
(800) 257-7742

Swarovski America Ltd.
2 Slater Road
Cranston, Rhode Island
02920
(401) 463-3000

Swift Instruments Inc.
952 Dorchester Avenue
Boston, Massachusetts
02125
(617) 436-2960

Tasco
P.O. Box 520080
Miami, Florida
33152
(305) 591-3670

Zeiss Optical Inc.
P.O. Box 2010
Petersburg, Virginia
23804
(804) 861-0033

Appendix B:

Tripod Manufacturers

A brand name different from that of the distributor or man-ufacturer is in parentheses.

Berkey Marketing Companies
  (Slik)
25–20 Brooklyn-Queens
  Expressway West
Woodside, New York
11377
(718) 932-4040

Bogen Photo Corporation
100 South Van Brunt Street,
P.O. Box 448,
Englewood, New Jersey
07631
(201) 568-7771

Davis & Sanford Company
24 Pleasant Street
New Rochelle, New York
10802
(914) 632-1636

HP Marketing (Linhof)
216 Little Falls Road
Cedar Grove, New Jersey
07009
(201) 808-9010

Karl Heitz Inc. (Gitzo)
34–11 62 Street, P.O. Box 427
Woodside, New York
11377
(718) 565-0004

E. Leitz Inc.
Link Drive
Rockleigh, New Jersey
07647
(201) 767-7500

Spiratone
135–06 Northern Boulevard
Flushing, New York
11354
(718) 886-2000

Uniphot/Levit Corporation
  (Star-D)
P.O. Box 429
Woodside, New York
11377
(718) 779-5700

Velbon International
  Corporation
2433 Moreton Street
Torrance, California
90505
(213) 530-5446

Vivitar Corporation
1630 Stewart Street
Santa Monica, California
90406
(213) 870-0181

Appendix C:

Selected Magazines

American Birds
950 Third Avenue
New York, New York
10022
(212) 546-9191

Astronomy
Kalmbach Publishing Company
21027 Crossroads Circle
P.O. Box 1612
Waukesha, Wisconsin
53187
(414) 796-8776

Audubon
    Society
Membership Data Center
P.O. Box 2666
Boulder, Colorado
80322
(800) 525-0643

Birder's World
720 East 8th Street
Holland, Michigan
49423
(616) 396-5618

Birding
American Birding Association,
  Inc.
P.O. Box 470
Sonoita, Arizona
85637

Bird Watcher's Digest
P.O. Box 110
Marietta, Ohio
45750
(800) 421-9764

British Birds
c/o Mrs. Erika Sharrock
Fountains, Park Lane
Blunham, Bedford, MK44 3NJ
England

Popular Photography
1 Park Avenue
New York, New York
10016

Sky & Telescope
P.O. Box 9111
Belmont, Massachusetts
02178–9111
(617) 864-7360

Wildbird
P.O. Box 6050
Mission Viejo, California
92690
(714) 855-8822

Glossary
of Optical Terms

aberration—a term combined with various adjectives to describe flaws in an optical system. Chromatic aberration, for example, refers to spurious color.

absorption—the process by which glass diminishes the amount of light passing through it.

catadioptric—an optical system utilizing both mirrors and lenses.

center focus—the term used when both eyepieces of a binocular are focused simultaneously from the center post that joins the binocular halves.

close focus—the ability of binoculars to focus on objects about 15 feet away or less.

coating—any of several chemical compounds applied to optics to reduce the amount of light scattered from their highly polished surfaces.

collimation—the precise alignment of all parts of an optical system. Accurate collimation is necessary to obtain the best image possible.

contrast—a measure of how well details in an image are discriminated from one another. The ability of an optical system to make visible adjacent features that have nearly equal brightness, colors, or both.

dioptre—a measure of the power (curvature) of a lens; these units are also used to indicate degrees of near- and farsightedness.

erector—an optical device that produces a right-side-up image.

exit pupil—the bright circle seen in an eyepiece when viewed from a distance of a few inches. The exit-pupil diameter is calculated by dividing the objective diameter by the magnification. Compare "pupil."

eyepiece—the part of a binocular or telescope that you look into. Eyepieces can have one "fixed" power or zoom continuously from low to high power.

eyepoint—the distance behind the eyepiece where the eye must be placed to see a full field of view.

eye relief—the distance between your eye and the eyepiece lens nearest your eye. The amount of eye relief is generally inconsequential for persons who do not require eyeglasses. Eyeglass wearers usually need 17 mm or more of eye relief. So-called "long-eye-relief" instruments are designed to accommodate eyeglass wearers.

field of view—the width of the image you see when looking through an optical instrument. Human eyes have a field of view of about 180 degrees; a typical binocular has a field of about 7 degrees. The fields cited in this book are technically known as "real" or "true" fields of view—they are what you see when you look through an instrument. The field of view varies according to the ratio of the focal length of the objective to the diameter of the eyepiece's so-called field lens (the one that faces the objective). There is also a so-called apparent field of view, which can be found approximately by multiplying the true angular field by the instrument's magnification. (For example, an 8-power binocular with a 7-degree true field of view has a 56-degree apparent field of view.) The larger the apparent field of view the more truly wide-angle an optical system is, regardless of its magnification.

focal length—the distance between a lens (or mirror) and a surface on which it projects a sharp image of an object at "infinity."

focal ratio—the focal length of an objective divided by the diameter of the objective.

individual focus—the term used when the two eyepieces of a binocular are focused separately.

interpupillary distance/scale—the separation between the centers of a person's eyes and the scale on binoculars calibrated to match this separation.

lens—an optical element that forms an image by refracting (bending) the path of light rays.

magnification—the amount the image size is increased by an optical system over that seen by the unaided eye; often called the "power" of the optical system.

mirror—an optical element that forms an image by reflecting light rays.

objective—the front lens or principal mirror of an optical instrument.

parfocal—an expression indicating that eyepieces having different magnifications can be interchanged without affecting precise focus. Parfocal also applies to the ability of zoom systems to stay in perfect focus.

Porro-prism—a prism that twice reflects light at right angles; two of these prisms are used in each half of a binocular.

pupil—the dark center of a person's eye that expands or contracts according to the prevailing light level. Compare "exit pupil."

resolution—the ability of an optical system (including the eye) to separate closely adjacent details. The resolution attained depends on optical quality as well as on the brightness and contrast of the subject.

roof prism—actually two prisms mounted in-line to allow the barrels of a binocular to have a compact, cylindrical contour.

scattering—any process that deflects light away from forming an image. Scattering produces lowered contrast.

twilight factor—an indication of how well a binocular will perform under low-light conditions. It is calculated by taking the square root of the quantity (magnification × objective diameter in millimeters).

wide-angle—for binoculars, this term is often used for instruments having real fields of view of roughly 8 degrees or more. Technically, wide-angle or wide-field generally applies to any optical system having an apparent field of view of 60 degrees or more.

zoom—the ability of an optical instrument to change magnification continuously throughout its design range.

Index